FROM
SALISBURY TO EXETER
The Branch Lines

FROM
SALISBURY TO
EXETER
The Branch Lines

Derek Phillips

OPC
Oxford Publishing Co

Front cover: Adams '415' class 4-4-2T No 30584 pulls away from Combpyne heading towards Lyme Regis on 8 July 1959. *R. C. Riley*

Rear cover, upper: Drummond M7 0-4-4T No 30676 at Exmouth on 13 October 1959. *R. C. Riley*

Rear cover, lower: 2-6-2T No 82013 arrives at Sidmouth with a train from Sidmouth Junction. *R. C. Riley*

Half title page: With the regulator closed and the injector turned on to keep the engine quiet in the station 'M7' No 30131 runs past Yeovil South Junction with the 6.52pm from Yeovil Junction on 16 August 1959. *Terry Gough*

Title page: The signalman at Exmouth signalbox waits on the veranda to collect the single-line token from the fireman of No 82023 arriving with a four-coach train from Exeter Central on 13 October 1959. The signalbox opened on 20 July 1924 replacing the original, and was closed on 10 March 1968. *R. C. Riley*

First published 2000

ISBN 0 86093 546 9

© Derek Phillips 2000

Published by Oxford Publishing Co

an imprint of Ian Allan Publishing Ltd, Terminal House, Shepperton, Surrey TW17 8AS.
Printed by Ian Allan Printing Ltd, Riverdene Business Park, Hersham, Surrey KT12 4RG.

Code: 0010/A2

Acknowledgements

I am very grateful for the advice, help and photographic assistance given to me by: Ben Ashworth, Hugh Ballantyne, R. M. Casserley, John Day, Terry Gough, the HMRS, George Pryer, Ron Lacey, Lens of Sutton, S. C. Nash, R. C. Riley, the Signalling Record Society, the South Western Circle, Adrian Vaughan, and Yeovil Library. Also, many thanks to Norman Pattenden for checking the manuscript and his kind advice on corrections and additions etc, and also to Mike King for delving through the files of the Wessex-Collection.

Bibliography

LSWR Locomotives: The Early Engines and the Beattie Classes, D. L. Bradley, Wild Swan, 1989.
LSWR Locomotives: The Drummond Classes, D. L. Bradley, Wild Swan, 1986.
LSWR Locomotives: The Urie Classes, D. Bradley, Wild Swan, 1987.
Locomotives at the Grouping 1: Southern Railway, H. C. Casserley and S. W. Johnston, Ian Allan, 1965.
Our Home Railways Volume 1, W. J. Gordon, Frederick Warne, 1910.
The Southern West of Salisbury, Terry Gough, OPC, 1984.
L&SWR Engine Sheds Western District, Chris Hawkins and George Reeve, Irwell Press, 1990.
LSWR Then and Now, Mac Hawkins, David & Charles, 1993.
Southern Railway Branch Lines in the Thirties, R. W. Kidner, Oakwood Press, 1976.
Somerset Railways, Robin Madge, Dovecot Press, 1984.
Branch Lines of Somerset, Colin G. Maggs, Alan Sutton, 1993.
Branch Line to Lyme Regis, V. Mitchell and K. Smith, Middleton Press, 1987.
Southern Steam, O. S. Nock, David & Charles, 1966.
Working Yeovil Steam, Derek Phillips, Fox & Co, 1989.
Working the Chard Branch, Derek Phillips and Ron Eaton-Lacey, Fox & Co, 1991.
The Salisbury to Exeter Line, Derek Phillips and George Pryer, OPC, 1997.
Signal Box Diagrams of the Great Western & Southern Railways Volume 5: Exeter Central-Templecombe & Branches, George Pryer, 1995.
Portfolios on the Branch Lines, The South Western Circle.

Contents

Introduction

Passengers heading for the West Country in the relative comfort of main line stock, complete with a restaurant car and hauled by a Bulleid Pacific from Waterloo, would alight from the train at a junction station and find awaiting them that well-loved, much photographed and now almost forgotten part of the railway system — the country branch train. An 'M7' or 'O2' class 0-4-4 tank locomotive with one, or even two, venerable coaches which had seen better days, would be waiting alongside the branch platform.

There is a gentle hissing of steam from the locomotive, followed by the scrape of the fireman's shovel as he prepares his fire; carriage doors slam shut with a solid clunk as the passengers climb aboard. The carriages are clean and have the musty aroma of many years of use. Old sepia photographs advertising holiday haunts of times past line the compartments. As the train travels along the branch line at a sedate pace, its old coaches creak and groan and the wheel flanges squeal in protest at the curvature of the track. It may well have been the same locomotive and even the same coaches which had met the passengers journeying to their holiday destination the year before.

The trains were also a lifeline to local people in rural areas but here, as in most country districts, that opponent of the railway — the motor omnibus — was becoming prevalent. Upon reaching Axminster the first of the branch lines to the coast was encountered with that most photographed of all locomotive types on the route from Salisbury to Exeter Central — the graceful Adams Radial tanks — steaming impatiently away in the bay platform while waiting to head off towards the Dorset resort of Lyme Regis. The intrepid enthusiast in the know would be found at Axminster on summer Saturdays when the sight of *two* Adams Radials would be a joy to behold, as one locomotive arrived from Exmouth Junction to replace the other for the coming week and assisted on the branch with the through Waterloo coaches that day.

In compiling the illustrations for this volume I decided to leave out views of soulless diesel units and trackless, destroyed stations, preferring instead to keep the images of the branch lines as everyone knew them. The only route still alive today and thriving is the line from Exeter Central to Exmouth, but through the pages of this book the long-closed lines to Yeovil Town, Chard, Lyme Regis, Seaton and Sidmouth are open again, to remind us, on our gridlocked roads of today, that there was once a gentler and more ambient way to travel — in the heyday of the branch line train.

Derek Phillips
Yeovil
Somerset

Above left: Adams Radial 4-4-2T No 30584, hauling a single LSWR coach, No S6401S, approaches Lyme Regis station with the 8.43am from Axminster on 3 July 1956.
Hugh Ballantyne

Left: 'West Country' class Pacific No 34104 *Bere Alston* climbs the 1 in 45 out of Tipton St John's with a Plymouth to Sidmouth excursion on 3 August 1959 with 2-6-2T No 82017 giving banking assistance at the rear. The line to Budleigh Salterton can be seen to the left. *S. C. Nash*

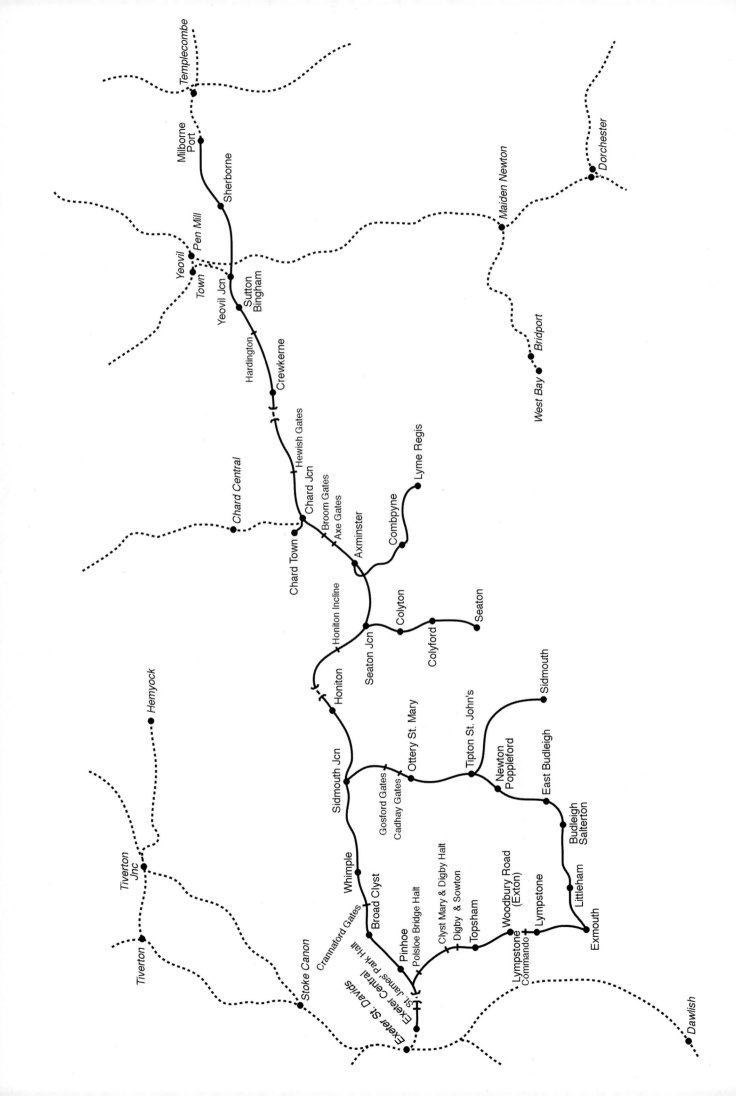

Above: A classic branch line scene. Lower quadrant signals and a small signalbox with the signalman standing by the trackside waiting to receive the single-line token as No 82010 arrives at Sidmouth on 13 October 1959. *R. C. Riley*

YEOVIL BRANCH.

(Distance 1 mile 65 chains.)

Not Saturdays, 18th July to 9th September inclusive, and daily commencing 12th September.

To Yeovil.

	WEEKDAYS.		Yeovil Junc. dep.	Yeovil arr.
a.m.	Engine	NM	2 10	2 15
"	"	NM	3 35	3 40
"	Pass. & Mail	NM	4 25	4 30
"	News & Mail }		4 55	5 0
"	Freight	...	7 0	7 5
"	Pass.	B	7 35	7 39
"	Pass.	BC	8 5	8 9
"	Pass.	B	8 29	8 33
"	Pass.	...	9 4	9 9
"	Pass.	B	9 43	9 47
"	Pass.	...	10 27	10 32
"	Pass.	B	10 55	10 59
"	Pass.	...	11 22	11 27
"	Pass.	B	11 50	11 54
p.m.	Pass.	B	12 15	12 19
"	Pass.	B	1 0	1 4
"	Freight	NS	1 15	1 20
"	Pass.	BD	1 53	1 57
"	Pass.	BP	2 3	2 7
"	Pass.	...	2 16	2 21
"	Pass.	...	3 18	3 23
"	Pass.	B	3 40	3 44
"	Pass.	FOBF	4 2	4 6
"	Pass.	B	4 23	4 27
"	Pass.	B	5 10	5 14
"	Pass.	B	5 33	5 37
"	Pass.	BE	5 47	5 51
"	Pass.	BF	6 0	6 4
"	Pass.	B	6 36	6 40
"	Pass.	B	7 7	7 11
"	Pass.	B	7 53	7 57
"	Engine	...	8 0	8 5
"	Pass.	...	8 32	8 37
"	2 Engines	...	8 45	8 50
"	Pass.	BNS	9 0	9 4
"	Pass.	BSO	9 3	9 7
"	Pass.	...	9 15	9 20
"	Pass.	B	9 33	9 37
"	Engine	...	10 2	10 7
"	Engine	NSA	11 0	11 5

From Yeovil.

	WEEKDAYS.		Yeovil dep.	Yeovil Junc. arr.
a.m.	Engine	NM	2 55	3 0
"	2 Engines	...	4 0	4 5
"	Engine	MO	4 50	4 55
"	Engine	H	5 20	5 25
"	Engine	...	6 10	6 15
"	Pass.	...	6 25	6 30
"	Pass.	B	6 45	6 49
"	Engine	MO	6 50	6 55
"	Pass.	...	7 0	7 5
"	Pass.	...	7 37	7 42
"	Pass.	B	7 46	7 51
"	Pass.	B	7 57	8 1
"	Engine	...	8 5	8 10
"	Pass.	B	8 18	8 22
"	Pass.	B	9 26	9 30
"	Engine	NS	9 45	9 50
"	Pass.	...	10 15	10 20
"	Pass.	BMO	10 30	10 34
"	Pass.	BNM	10 38	10 42
"	Pass.	...	11 8	11 13
"	Pass.	B	11 35	11 39
"	Freight	NS	11 45	11 50
p.m.	Pass.	B	12 2	12 6
"	Pass.	B	12 47	12 51
"	Pass.	BD	1 35	1 39
"	Pass.	BP	1 48	1 52
"	Freight	SO	1 55	2 0
"	Engine	...	2 20	2 25
"	Pass.	...	2 30	2 35
"	Pass.	B	3 18	3 22
"	Pass.	B	3 50	3 54
"	Engine	...	4 0	4 4
"	Pass.	R	4 R10	4 15
"	Pass.	FOBF	4 16	4 20
"	Engine	...	4 50	4 55
"	Pass.	B	4 58	5 2
"	Pass.	B	5 22	5 26
"	Engine	...	5 30	5 35
"	Pass.	B	5 40	5 44
"	Freight	NS	5 52	5 57
"	Engine	...	6 0	6 5
"	Pass.	B	6 15	6 19
"	Vans	...	6 40	6 45
"	Pass.	B	6 50	6 54
"	Engine	NSA	7 20	7 25
"	Pass.	B	7 40	7 44
"	Pass.	B	8 45	8 49
"	Pass. & Mails	B	9 22	9 26
"	Engine	SO	10 30	10 35

Saturdays only, 23rd July to 10th September inclusive.

To Yeovil.

			Yeovil Junc. dep.	Yeovil arr.
a.m.	Engine	...	2 10	2 15
"	"	...	3 35	3 40
"	Pass. & Mail	...	4 25	4 30
"	News & Mail }		4 55	5 0
"	Freight	...	7 0	7 5
"	Pass.	B	7 35	7 39
"	Pass.	BC	8 5	8 9
"	Pass.	B	8 29	8 33
"	Pass.	...	9 4	9 9
"	Pass.	B	9 43	9 47
"	Pass.	...	10 27	10 32
"	Pass.	B	11 0	11 4
p.m.	Pass.	B	12 10	12 14
"	Pass.	B	12 27	12 31
"	Pass.	...	12 41	12 46
"	Pass.	B	2 0	2 4
"	Pass.	B	3 18	3 22
"	Pass.	B	3 40	3 44
"	Pass.	B	4 2	4 6
"	Pass.	B	4 23	4 27
"	Pass	B	5 10	5 14
"	Pass.	B	5 35	5 39
"	Pass.	B	6 3	6 7
"	Pass.	B	6 36	6 40
"	Pass.	B	7 7	7 11
"	Pass.	B	7 53	7 57
"	Engine	...	8 0	8 5
"	Pass.	...	8 38	8 43
"	2 Engines	...	8 45	8 50
"	Pass.	B	9 3	9 7
"	Pass,	...	9 15	9 20
"	Pass.	B	9 42	9 46
"	Engine	...	10 2	10 7

From Yeovil.

			Yeovil dep.	Yeovil Junc. arr.
a.m.	Engine	...	2 55	3 0
"	2 Engines	...	4 0	4 5
"	2 Engines	...	5 20	5 25
"	Engine	...	6 10	6 15
"	Pass.	...	6 25	6 30
"	Pass.	B	6 45	6 49
"	Pass.	...	7 0	7 5
"	Pass.	...	7 37	7 42
"	Pass.	...	7 46	7 51
"	Pass.	B	7 57	8 1
"	Engine	...	8 5	8 10
"	Pass.	B	8 18	8 22
"	Engine	...	9 20	9 25
"	Pass.	B	9 26	9 30
"	Pass.	...	10 15	10 20
"	Pass.	B	10 46	10 50
"	Pass.	...	11 50	11 55
"	Pass.	B	11 56	12 0
p.m.	Pass.	B	12 16	12 20
"	Pass.	B	1 45	1 49
"	Freight	...	1 50	1 55
"	Engine	...	2 15	2 20
"	Pass.	...	2 25	2 30
"	Scouts	BQ	2 45	2 49
"	Pass.	B	3 5	3 9
"	Pass.	B	3 27	3 31
"	Engine	...	3 45	3 49
"	Pass.	B	3 50	3 54
"	Pass.	...	4 5	4 10
"	Pass.	B	4 16	4 20
"	Engine	...	4 50	4 55
"	Pass.	B	4 58	5 2
"	Pass.	B	5 22	5 26
"	Engine	...	5 30	5 35
"	Pass.	B	5 50	5 54
"	Engine	...	6 0	6 5
"	Pass.	B	6 15	6 19
"	Vans	...	6 40	6 45
"	Pass.	B	6 50	6 54
"	Pass.	B	7 40	7 44
"	Pass.	B	8 45	8 49
"	Pass. & Mails	B	9 25	9 29
"	Engine	...	10 30	10 35

A—Works a Freight or Cattle Train if required. B—Branch Train. C—Extra vehicle traffic, except for G.W.R. via Yeovil (Pen Mill), must not be attached to this train. D—Not Saturdays 18th July to 9th September only. E—Mondays to Thursdays only from 18th July to 8th September, inclusive: daily commencing 12th September. F—Fridays from 22nd July to 9th September only. H—One engine M.O. two other days. P—Commencing 12th September. R—On Fridays 22nd July to 9th September only runs five minutes earlier.

YEOVIL (TN.) & PEN MILL (G.W.R.)

(Distance 47 chains.)

To Pen Mill.

	WEEKDAYS		Yeovil (Town). dep.	Pen Mill. arr.
a.m.	Pass.	BA	10 10	10 12
p.m.	Pass.	A	12 55	12 57
"	Pass.	BA	8 5	8 7

From Pen Mill.

	WEEKDAYS		Pen Mill. dep.	Yeovil (Town). arr.
a.m.	Pass.	BA	10 15	10 17
p.m.	Pass.	A	1 18	1 20
"	Pass.	BA	8 30	8 32

A—Worked by S. Rly. Company. B—Yeovil Branch Train.

Above: The Southern Railway Working Timetable for July 1923 regarding the Yeovil branch shows the intensity of traffic using this busy double-track line, including light engines, passenger and freight traffic.

Yeovil Town-Yeovil Junction

The Somerset Borough of Yeovil, home of Westland Helicopters, St Ivel butter and the most famous giant killing football team of them all, was once a major manufacturing centre in the glovemaking industry. John Leland, in 1540, called Yeovil 'a good market town', and Collinson, writing in 1791, noted that 'glovemaking had become the principal manufacture'. Glovemaking in the town can be traced back to the 14th century and this was still the case in 1856 with the then largest glove manufacturers in the town, Messrs Boyd & Fook, paying wages totalling £7,000 per annum. Passenger travel to and from the town, situated on the route from London to the West, before the coming of the railway, was handled by stage coaches which ran from various hostelries such as the Three Choughs, the Mermaid and the Castle Inn, to London, Exeter, Bristol, Bath, Weymouth and Taunton. By 1840 the *Quicksilver Mail, Traveller* and *Royal Mail* would complete the journey to the Brown Bear, Piccadilly, London, in one day, and in 1850 the *Telegraph* would leave the Mermaid Inn at 7.30am for Salisbury where passengers continued their journey to London by train. The Castle Inn also offered services to Bath and Bristol by *Edwards Omnibus*, while the Three Choughs Hotel ran services to Dorchester by the *Royal Mail*. However, the Bristol & Exeter Railway sounded the death knell for the stagecoach with the opening from Durston, situated on the Bristol-Taunton main line, of a 24-mile single-track branch line to a terminus situated on the western outskirts of Yeovil, at Hendford. Constructed to the broad gauge under the Act of 31 July 1845, the line

Below: An early aerial view of Yeovil Town shows the impressive overall roof of the station to the right. The goods shed is in mid-centre with the gasworks in the background. The yard is packed full of wagons and vans. Yeovil Town West signalbox can be seen at the bottom left-hand corner to the right of the footbridge. This box opened in 1882 and was replaced in 1916. *Lens of Sutton*

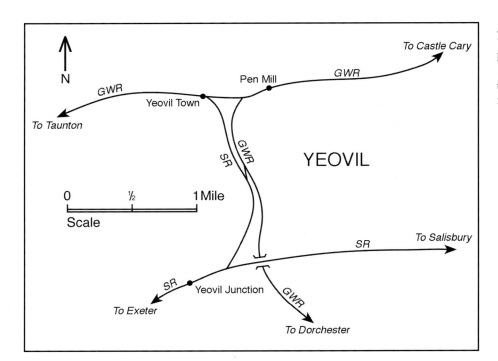

opened to passenger services on 1 October 1853 and
to goods traffic on 26 October the same year.

Hendford, described as a quiet leafy glade before the
coming of the railway, was one of the ancient tythings
of the town belonging to the Maltravers family —
Lords of the Manor of Yeovil. The first owner, Hugh
Maltravers, was witness to King Henry I's charter to
the monks of Montacute. Facilities at Hendford at the
time of the opening of the line comprised a single
storey stone-built station building with a locomotive
shed and a goods shed nearby. Traffic growth was
phenomenal: passenger traffic receipts from Yeovil
had reached a total of 30,000 in 1855, and by 1856, six
passenger trains arrived and departed each weekday
(with two on Sundays) in addition to several luggage
trains and three goods trains.

The Great Western Railway, by contrast, appeared
on the north-east edge of Yeovil on 1 September 1856
with the opening of Pen Mill station. The 26 miles of
the Wilts, Somerset & Weymouth broad gauge main
line from Frome was built as single track, although all
the earthworks were made wide enough for double
track. The line was extended to Weymouth on
20 January 1857 and the Bristol & Exeter lost no time
in completing a single line connecting spur from
Hendford to the GWR at Pen Mill, this link opening
on 2 February 1857. The third, and arguably the most
important railway company to reach Yeovil, the
standard gauge London & South Western Railway,
was now set to appear. The main line from Salisbury,
constructed by the nominally independent Salisbury
& Yeovil Railway under the Act of 21 July 1856 and
operated by the LSWR, had opened in stages.
Salisbury to Gillingham on 2 May 1859, Gillingham
to Sherborne on 7 May 1860, and finally the section
from Sherborne to Yeovil (Hendford) via the Bradford
Abbas cutting, on 1 June 1860. This allowed LSWR
trains to run direct to the Bristol & Exeter station at
Hendford. The S&Y track crossed the GWR's

Weymouth line by an overbridge and ran alongside
before turning west and running parallel to the broad
gauge B&E Hendford to Pen Mill spur from the site
of the future Town station to Hendford. The B&E had
laid this section of independent standard gauge track
for approximately one mile alongside their own
permanent way. The gauge was also mixed in the yard
at Hendford for the South Western locomotives and
stock, although relations were not that harmonious as,
in May 1860, the B&E objected to the standard gauge
engines using its shed and turntable. This prompted
the LSWR to move the turntable temporarily from
Sherborne to Yeovil for the opening of the line, with
its initial services comprising five trains from Yeovil
and four from Salisbury.

Yeovil Junction and the western extension of the
LSWR main line to Exeter Queen Street opened a few
weeks later on 19 July 1860, the link from Yeovil
Junction to Yeovil Upper Junction now giving trains
from Yeovil a direct link to Exeter. The main line
services were then worked via Yeovil Junction, thus
avoiding the Bradford Abbas cutting and the direct
line to Hendford. The short connecting link was
constructed between the Salisbury-Yeovil and the
Yeovil-Exeter line, but at least one main line service
still ran via Yeovil Junction-Hendford-Sherborne
through the Bradford Abbas cutting. A connecting
service was also run between Hendford and Yeovil
Junction, the branch between Yeovil and Yeovil
Junction being doubled in 1864. With the realisation
that the cramped site at Hendford would not be ideal
to serve the town following the arrival of the line from
Salisbury and the extra traffic it would bring, the
Salisbury & Yeovil Railway began negotiations for a
joint station with the Bristol & Exeter. This was
without the knowledge of the LSWR in August 1858.
However, the South Western after protestations,
swallowed its pride and participation was
provisionally agreed, despite bitter complaints from

the S&Y over the dallying of the LSWR regarding approval of the plans, which resulted in serious reduction of the S&Y's revenue by non-completion of the station. Mention of the new joint station is found in the S&Y half-yearly report of July 1859 in which Joseph Locke announced that agreement had been reached with the B&E for the construction of a station and that surveys were in progress. Louis Ruegg, a prominent shareholder in the Salisbury & Yeovil and editor of the *Sherborne Journal*, commented in his newspaper that the site was not fixed, but it would be at the back of the 'gashouse'. The gasworks positioned off Middle Street predated the arrival of the railway in the town and was built by the Yeovil Gas & Coke Company in 1833 on the site of a drained withy bed. By 1856 it was producing 28,000 cubic feet of gas per day and supplying 124 of the corporation's street lamps and became rail connected with the opening of the station goods yard.

On 30 July 1859 Locke, Brassey and Townsend, the latter possibly a resident engineer, travelled by train to Gillingham, at that point in time the railhead (the line from there to Sherborne not opening until 7 May 1860), and thence by coach and four to Yeovil, to set out the site of the new station. After being met by the local engineers Fraser and Harrison, they examined the area and decided that the site would be on a field between the gasworks and Newton Road. I often wonder if the good townsfolk of Yeovil knew that the eminent and respected figures of Joseph Locke and Thomas Brassey were in their midst on that far off day in 1859. Joseph Locke, born in 1805, became articled to George Stephenson at Newcastle in 1823 and worked with the great man on the Liverpool &

Manchester Railway until a disagreement with Stephenson in 1832 led to Locke's resignation from the L&M. He then set up on his own account as a civil engineer and built the Grand Junction Railway from Birmingham to Warrington, 1835-37; the London & Southampton 1836-40; Sheffield & Manchester 1838-40; Lancaster & Preston Junction 1837-40 and the Lancaster & Carlisle 1843-6, as well as lines in Scotland, including the Scottish Central and the Caledonian from Carlisle to Glasgow and Edinburgh. He was also responsible for various railways in Europe including Paris-Rouen; Rouen-Havre and Barcelona-Mataro. Locke was elected Liberal MP for Honiton in 1847. He died on 18 September 1860 at the age of 55.

Born in Buerton, Cheshire, in 1805, Thomas Brassey, the great civil engineer and contractor, ranks high with Locke in the history of our railways. In 1835, Locke awarded him the contract for the construction of 10 miles of the Grand Junction following his building of Penkridge Viaduct. He worked as contractor to Joseph Locke on many projects in Great Britain, including the London & Southampton Railway on which he employed 3,000 men, and also on railway contracts in Norway, Sweden, Switzerland, Turkey, India, Australia and South America. At one time he had 75,000 men in his employment and his weekly payments for labour amounted to £15,000-£20,000 with a capital of £36 million. No wonder he was known as the 'navvy king'. Always known for his scrupulous honesty and fairness, and well respected for his ability to select the right men for the job, Thomas Brassey died on 8 December 1870 aged 65.

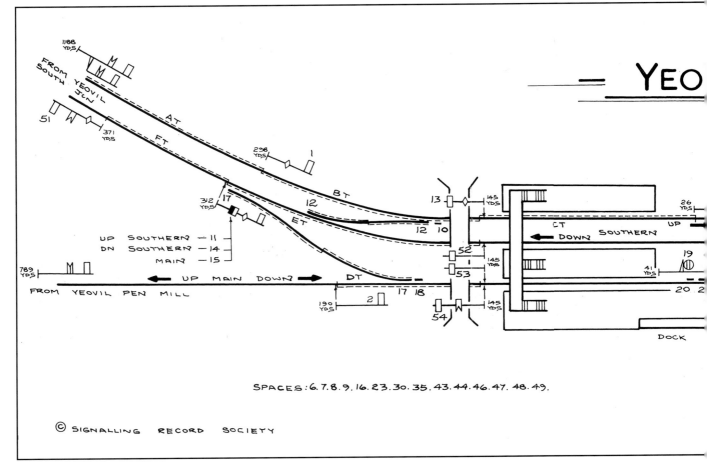

SPACES: 6.7.8.9.16.23.30.35.43.44.46.47.48.49.

© SIGNALLING RECORD SOCIETY

Towards the end of August 1859 the directors of the LSWR visited Yeovil to decide whether the goods depot should be at the Stoford passenger station (Yeovil Junction), or by Yeovil Gashouse. Some doubt seems to have remained regarding the intended site of the joint station, for the *Sherborne Journal* in March 1860 confirmed it was to be near the Elephant & Castle in Sherborne Road. A Bill was drafted to authorise the construction of the station, and at a meeting of LSWR shareholders on 3 May 1860 presided over by Charles Castleman, the deputy chairman, approval was given, including agreement of expenditure to the sum of £20,000. On the same day, representatives of the railway companies involved met at Yeovil to discuss the arrangements and it was decided that construction of the new station should be carried out by the LSWR. The Bill received Royal Assent on 14 June 1860, 'An Act for the alteration at Yeovil of the Salisbury and Yeovil Railway leased to the London and South Western Railway Company of the Yeovil and Durston Branch of the Bristol and Exeter Railway Companies; and for other purposes.' Work began soon after the opening of the Salisbury line, and by October 1860 it was reported that 'building is going on in right earnest, and in a short time there will doubtless be a handsome and commodious building.' Despite the employment of over 100 men, the station was not ready until June 1861, but there was no opening ceremony, quite unlike the celebrations that greeted the arrival of the Bristol & Exeter at Hendford in 1853.

Yeovil Town

The joint station at Yeovil Town — the most important and convenient of the three stations for the local populace — opened on 1 June 1861. Hendford was then relegated to goods traffic and the station building used for stabling horses. The handsome and grand station façade with projecting wings and a large central gable was designed in the Tudor style by Sir William Tite (architect of the Royal Exchange and the original Nine Elms terminus, and many other station buildings on the Salisbury-Exeter route). It was constructed in red brick with creamy ashlar dressings, being perhaps the most important commercial building to appear in the town. The three platforms were provided with two overall roofs containing some 15,000 square feet of glass and supported by iron girders and pillars. Both railway companies now used the new station with its improved facilities; a central waiting room was flanked by two ladies' waiting rooms, other offices on the 450ft-long No 1 platform consisting of the LSWR telegraph office, booking office, parcels office, refreshment room, Wymans bookstall, two gentlemen's lavatories and a lamp room. The GWR in the manner of a joint station proper also had its own booking office, parcels office and porters' room. No 1 platform was used by the Durston to Yeovil trains, and a bay platform to the rear of this provided accommodation for parcels and milk churn traffic etc.

TOWN ——

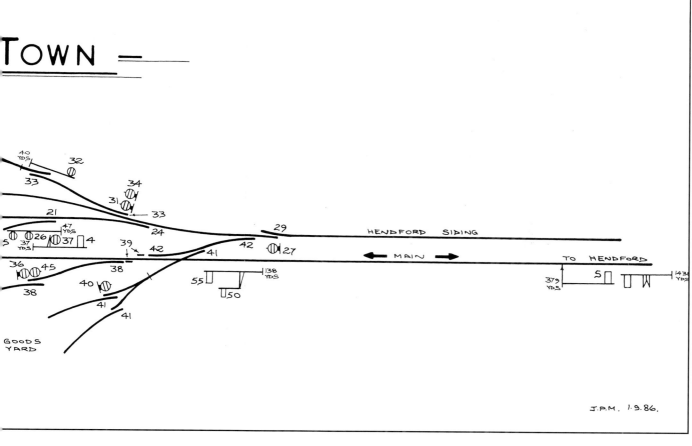

J.P.M. 1.9.86.

No 2 platform used by the LSWR outgoing services was of the island pattern, 330ft long and 15ft wide. It was completely devoid of buildings of any kind and was specially designed for the transfer of passengers from Durston branch trains to the LSWR's services. No 3 platform, which backed on to the locomotive shed yard and its associated sidings, was the same length as its opposite number but was narrower and, again, no buildings of any kind were provided. It was used by incoming services from Yeovil Junction. A footbridge connected all three platforms. A large building, situated near the platform and used by the guards, contained on the ground floor a lofty dining hall fitted with a cooking range, gas cooking appliances, lockers and lavatories etc. The upper floor was for use by the men who worked the night goods trains between London and Yeovil and had two bedrooms with three beds in each room.

All lines and sidings were laid to mixed gauge. The B&E and LSWR had separate departments for all services, except for signalling which was under the control of the B&E and subsequently the GWR which absorbed the B&E on 1 January 1876. It was the policy of the LSWR not to operate its own refreshment rooms but to let them to individual tenants. The company received so many complaints regarding the variable quality of service, food, drinks, etc, that it was decided, from June 1882, to let all the refreshment rooms on its system to one contractor, Messrs Simmonds. From 1 January 1888 this changed to the well-known firm of Spiers & Pond for a sum of

£12,500 per year plus £100 for the Yeovil Town refreshment rooms.

Town station was not run as a true joint station until 30 March 1882, by which time the broad gauge had disappeared. Four signalboxes (GWR and LSWR each with a box at the east and west ends of the station) originally controlled the layout. The LSWR worked its goods line to Hendford on the train staff system quite independently of the parallel GWR line, this practice ceasing in 1879 when the former was worked as a long siding between Yeovil Town and Hendford. In 1867 the B&E was alarmed by the plans of the Somerset & Dorset Railway to reach Bridgwater and, in order to defeat this scheme, laid a third rail to its own main line (under the powers of 28 June 1866) from Highbridge through Bridgwater to Durston, and from there along the branch to the joint station at Yeovil for a total cost of £125,000. From November 1867 a daily 'narrow gauge' goods train ran between Yeovil and Bridgwater, the B&E complaining bitterly in September 1868 when the LSWR refused to allow passage of its new 'narrow gauge' stock through Town station, thus causing losses of £6,000 yearly while the carriages languished unused at Highbridge. The GWR laid mixed gauge track between Town and Pen Mill stations on 12 November 1868 and, at last, the B&E ran its 'narrow gauge' trains *through* Town station. The GWR's Weymouth line through Yeovil Pen Mill was converted from broad to standard gauge in June 1874.

Above: A works photograph taken by the Gloucester Railway Carriage & Wagon Company Ltd in September 1886 of coal wagon No 4 for the Yeovil Gas & Coke Co. A 10-ton six-plank open wagon fitted with spoked wheels and dumb buffers, it was painted black with white lettering. *HMRS ACG201*

With the GWR branch from Durston being converted to standard gauge on 30 June 1879, this brought to an end the costly and time-consuming need to tranship goods at Hendford between the two systems. Standardisation also brought relief from locomotives running round, as GWR locomotives could utilise the South Western tracks at Town station and vice versa. The GWR and LSWR jointly agreed on 23 February 1881 that the station should be equipped with modern signalling at an estimated cost of £1,536, the sum being shared between the two companies. From 30 March 1882 two new boxes, bearing the suffixes East and West, replaced the four original boxes. The 24-lever East box, situated in the fork of the lines to Pen Mill and Yeovil Junction, controlled the running road of the Durston branch from Yeovil Town West to the South box at Yeovil Pen Mill (GWR), plus all the shunting on the east side of the station and the LSWR trains to and from Yeovil Junction. The 37-lever West box was positioned beside the Durston branch line near the entrance points to the goods shed. Standing in the lee of the public footbridge leading from Stars Lane to Summerhouse Hill, it was responsible for controlling the single-track branch to Hendford, plus traffic in and out of the LSWR goods yard and locomotive shed, and also shunting operations of both companies' trains at the west end of the station. Many of the levers were slotted with those in the East box, the electric train staff being used between Yeovil Town East and Yeovil Pen Mill South; trains between the East box and Yeovil Junction were worked under the LSWR

double-line block, with the GWR electric train staff used between Yeovil West and Hendford. The station was an interesting area to control, with the double-track LSWR branch terminating at the station and the through GWR single line. There were four different sets of block instruments in the two cabins. Nos 1 and 2 roads were worked under GWR double-line rules, and the single line No 1 road by GWR single line block instruments. Special codes were used for trains and engines running through the station in the wrong direction and for the backing of coaches down the running roads.

Three sets of railway workers were employed at the station: LSWR, GWR (both with their own stationmaster) and joint staff. Each member of staff attended to the traffic of his own railway and also assisted with the transfer of traffic from one to another under certain regulations. The joint staff were responsible for the work of both companies, but were principally engaged in the movement of trains, and consisted of five signalmen, a lamp lad and a waiting room attendant. The signalmen for the East box were usually appointed by the LSWR and those for the West box by the GWR but both sets of men were conversant with the workings of each box and could undertake relief in either cabin. The LSWR stationmaster in 1909 was responsible for 45 staff including guards and joint personnel, while the LSWR engineers' staff consisted of a ganger plus six men. A ballast train stationed at Yeovil was manned by a foreman and 13 staff. The Great Western was responsible for the permanent way, with the LSWR

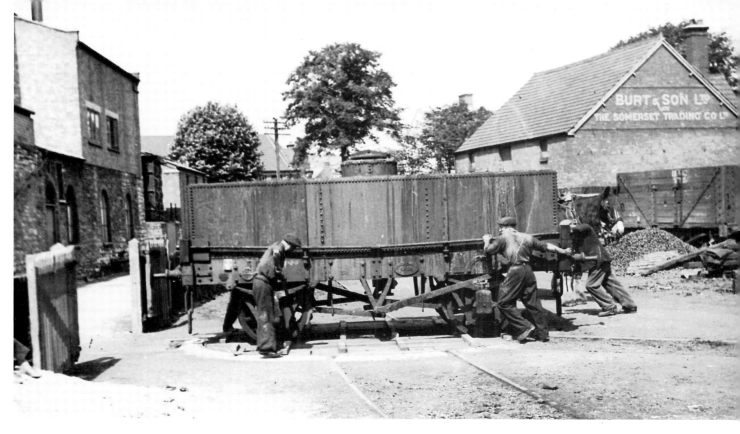

Above: The railway shunting horse looks on nonchalantly as 'human power' prepares to push a private owner tar tank off the wagon turntable in Yeovil Town goods yard and across Dodham Lane into the gasworks to the left, on 13 August 1946. The railway horses used for shunting were stabled near the goods shed, eventually being replaced by a diesel tractor fitted with a buffer beam. The building on the right is labelled Burt & Son Ltd late The Somerset Trading Co Ltd. *A. E. West*

Below: One of the Adams general purpose Class O4 0-4-2 locomotives, No E546, awaits the 'right away' with a train for Salisbury (with reversal at Yeovil Junction) at Yeovil Town on 2 August 1928. A total of 90 of these locomotives, known as the 'Jubilees', were built between 1887 and 1895. The overall roof covered more of the GWR part of the station than the Southern Railway side, which can be seen to good effect in this view. *H. C. Casserley*

Above: Class C8 4-4-0 No E292 from Salisbury shed stands in the Loco yard at Yeovil Town on 2 August 1928. The 'C8s' were the first four-coupled express engines built by Drummond for the LSWR. This particular engine had been rebuilt with a Urie smokebox door, boiler barrel clackboxes, sandboxes resited between the frames and a 4,000-gallon double-bogie tender. Dating from June 1898, the locomotive was withdrawn in January 1936. Nos 290-4 were allocated to Yeovil from Basingstoke in 1926 but all had been reallocated by the summer of the same year. The gasworks and part of the station overall roof can be seen in the background. *H. C. Casserley*

Below: GWR 2-6-2T No 4595 leaves Yeovil Town and takes the GWR single-line branch to Pen Mill station with the 11am from Taunton on 21 May 1935. *H. C. Casserley*

Above: Class U 2-6-0 No 1790 stands on the table road alongside Dodham Brook at Yeovil Town shed on 21 May 1935. This locomotive was the original 2-6-4T built by Richard Maunsell for the SE&CR as No 790 in 1917. Maunsell became CME of the new Southern Railway on 1 January 1923. Nine 2-6-4T engines were added to the class in 1925 (Nos A791-9) built by Armstrong Whitworth. Additional class members were added in 1926 with Nos A800-9 emerging from Brighton Works. All were named after rivers, the first becoming No A790 *River Avon*. After the fatal accident at Sevenoaks on 24 August 1927 when No A800 was derailed at speed, the entire class were rebuilt as 2-6-0 tender locomotives. Nos 1790-5 were reallocated to Yeovil from Eastbourne in 1933 and stayed for over 30 years.
H. C. Casserley

Below: Drummond 'L11' class 4-4-0 No E438 stands at the end of the coal road while being hand-coaled at Yeovil Town Loco on 12 June 1926. Fellow class member No E170 stands on the left. The locomotives were known by footplatemen as the 'Large Hoppers'. *H. C. Casserley*

Above: One of the famous Drummond 'Greyhound' 'T9s', No 30706, stands on the goods shed siding at Yeovil Town on 19 September 1952. The locomotive is in early British Railways lined black livery with plain tender sides. The double-bogie tender contains 4,000 gallons of water. No 706 was one of the first batch built by Dubs & Co of Glasgow. Dating from January 1899, the engine was withdrawn in May 1959. *A. E. West*

attending to repairs to the station building, including the portions used by the GWR. A large parcel and newspaper traffic was dealt with, as were goods from the nearby creamery of Aplin & Barrett, famous for the St Ivel brand of butter and cream. Deliveries for the LSWR around the town and nearby villages were handled by the horse vans of Messrs Chaplins, the GWR, of course, being responsible for its own cartage of goods from Hendford which was the principal goods depot for that company in Yeovil, and remaining as such for the town into Nationalisation until the ultimate closure of the yard.

A series of joint projects between the LSWR and GWR (under the agreement of 1910 to secure operating economies in areas served by both companies) to seek manpower and other savings during the constraints of wartime working resulted in the agreement of 27 January 1916 to combine the East and West signalboxes at Yeovil Town. Consequently, the two boxes were closed and replaced by a GWR brick-built box with a hip gabled roof which opened on 15 October 1916. The new box measured 33ft 6in x 11ft and was elevated 9ft, containing a 55-lever frame and positioned alongside the Durston branch at the west end of the station. The transfer of passengers between the two companies was quite affable and in 1909, for example, the Great Western Railway's 4.18pm to Taunton picked up passengers from the LSWR's connection off the 1pm from Waterloo, while the 4.25pm Yeovil-Portsmouth would be awaiting

passengers off the 4.18pm to Taunton. Other services, such as the 9.42am from Taunton, connected with the 11.10am Yeovil Town to Salisbury (arr 12.26pm) or Waterloo (arr 2.30pm), while the 7.30am from Weymouth with passengers changing at Pen Mill and Town stations would connect with the 9am Yeovil Town-Exeter (arr 10.9am) and Plymouth North Road (arr 11.51am). This pattern remained the norm, with exceptions, for the rest of the station's life. With the danger to passengers and staff alike of falling glass, the overall roofs were removed in 1934 and replaced by platform canopies.

The goods yard was at right angles to the station abutting Middle Street and flanked by the long approach road to the station frontage. Sidings served the local coal merchants, one trader being the Somerset Trading Company which, by 1946, was trading under the well-known local name of Burt & Son Ltd, and a wool store (later becoming Farr's scrapyard). A siding also served the gasworks, coal being shunted in and coke and tar oil shunted out. Locomotives were not permitted into the yard, wagons being moved through the goods shed at first by horses and, in later years, by a tractor equipped with a buffer beam. Wagon turntables were provided, and the goods shed (219ft long and 33ft wide) contained a 3-ton crane. Nearby, there were three adjacent unloading docks for the loading of horses, cattle and vehicles, etc and a 10-ton crane was also provided.

1934 Southern Railway Working Timetable... Yeovil Corporation Gas Works Siding — '

Connection to this private siding is provided from one of the Company's sidings by means of a turntable. Immediately beyond the gate near the Company's boundary the siding crosses the public cart road known as Dodham Lane. The point of exchange for vehicles to and from the private siding is on the Company's siding near the turntable and wagons are passed over the turntable and hauled to and from the Gas Works by the Company employees, who are responsible for the protection of road vehicles using the crossing during the time vehicles are passing to and from the siding.

1934 Southern Railway Working Timetable... Working of Pull and Push Trains Without a Guard...

A porter will accompany the train between Yeovil Junction and Yeovil Town when necessary for the purpose of giving assistance with the loading of parcels, luggage and mails. He must ride in the brake compartment and take charge of and sort mails, parcels and luggage. He will undertake no responsibility appertaining to the actual running of the train.

Left: Drummond 'K10' class 4-4-0 No 152 has just come over the crossover and on to the GWR line at Yeovil Town on 15 September 1948 in preparation to enter the goods shed siding. The six-wheeled tender contains 3,500 gallons of water. This class of locomotive was known as the 'Small Hoppers' in deference to the 'L11s', the 'Large Hoppers'. This particular locomotive, dating from December 1902, was withdrawn in February 1949. The large building in the centre background is the coking plant for the gasworks. *A. E. West*

Above: Class K10 4-4-0 No 389 shunts near the goods shed at Yeovil Town on 10 June 1949. Locomotives were not permitted in the shed, wagons being propelled through to the yard which lay at right angles behind the goods shed. The industrial area around the goods yard was dominated by the gasworks. Other users included Farr's scrapyard, which was also rail connected. The air would be heavy with the pungent aroma from the gasworks, the smell of animal skins from nearby tanneries, and smoke and steam from the station area. *A. E. West*

Station Road was also used as a parking area for Southern National buses. There was a time when Yeovil had one of the largest bus fleets in the South West for a town of its size, with Southern National having a large bus depot at Reckleford and an office at the corner of Middle Street and the station approach. In 1936, the GWR introduced a nonstop (albeit having to stop to exchange the single-line staff) diesel railcar service on the Yeovil to Taunton workings, the outward journey to Taunton starting at Weymouth with the return workings continuing as a semi-fast service to Trowbridge and Bristol Temple Meads via Yeovil Pen Mill. The diesel car timings to Taunton saved 20 minutes compared with the steam service. Passenger services were withdrawn between Yeovil and Durston on 15 June 1964, with freight services to Taunton being withdrawn from 6 July. Passenger trains ceased between Yeovil Town and Yeovil Junction on 2 October 1966, with a new service running between Yeovil Pen Mill and Yeovil Junction via the wartime connection at Yeovil South Junction. Town station remained open for parcels only until 1 March 1967, with Yeovil Town signalbox closing at the same time. Hendford Goods closed on 9 October 1967, the depot then being served by road transport. The track between Yeovil Town and Yeovil South Junction was taken out of use on 1 March 1967 and South Junction box closed a year later on 26 May, while the link from Pen Mill to the former Town station (including the remaining stub of the old B&E branch) which only remained in use for stabling at the old depot and for storage of track materials, succumbed in 1968. Town station site was taken over by the local authority and completely destroyed, a sad indictment of a society bereft of foresight, especially when the site was turned into a car park.

Yeovil Town Train Movements 1961 (Saturday Summer Service)

Train	Arr	Loco	Diagram	Dep	Destination
03.57 Junction	04.02	WC 4-6-2	Ex-Jcn 547		Loco to MPD
01.50 Eastleigh	04.29	BR4 2-6-0	Sal 48		Berth stock
		BR4 2-6-0	Sal 480	05.07	Light to Junction
03.30 Salisbury	05.16	BR4 2-6-0	Sal 478		Loco Stn Pilot
Loco ex-MPD		WC 4-6-2	Ex-Jcn 547	06.25	Ilfracombe
Loco ex-MPD	06.30	WC 4-6-2	Ex-Jcn 549		Carriage Pilot
		BR4 2-6-0	Sal 478	06.35	Light to Jcn
Loco ex-MPD		S15 4-6-0	Ex-Jcn 546	06.55	Waterloo
07.05 Pen Mill	07.07	5700 0-6-0PT	Yeovil 518	07.10	Taunton
Loco ex-MPD		M7 0-4-4T	Yeovil 517	07.22	Junction
		WC 4-6-2	Ex-Jcn 549	07.45	Ilfracombe
06.45 Taunton	07.48	45xx 2-6-2T	Taunton 60	07.50	Pen Mill
07.42 Junction	07.46	M7 0-4-4T	Yeovil 517	07.58	Junction
	08.11	M7 0-4-4T	Yeovil 517	08.20	Junction
08.20 Jcn Milk	08.25	WC 4-6-2	Sal 503		Berth stock
	08.39	M7 0-4-4T	Yeovil 517	08.44	Pen Mill
08.50 Pen Mill	08.52	M7 0-4-4T	Yeovil 517	09.08	Junction
09.16 Pen Mill	09.18	5700 0-6-0PT	Yeovil 519		Loco Stn Pilot
	09.25	M7 0-4-4T	Yeovil 517		
09.54 Pen Mill	09.56	45xx 2-6-2T	Taunton 60	09.57	Taunton
		M7 0-4-4T	Yeovil 517	10.00	Junction
10.20 Junction	10.24	M7 0-4-4T	Yeovil 517		
		M7 0-4-0T	Yeovil 517	10.50	Junction
10.05 Taunton	11.02	5700 0-6-0PT	Yeovil 518	11.05	Pen Mill
11.01 Junction	11.05	M7 0-4-4T	Yeovil 517	11.20	Junction
11.42 Junction	11.46	M7 0-4-4T	Yeovil 517	11.52	Junction
12.09 Junction	12.13	M7 0-4-4T	Yeovil 517	12.22	Junction
12.30 Junction	12.34	M7 0-4-4T	Yeovil 517		Loco to MPD
12.38 Pen Mill	12.40	5700 0-6-0PT	Yeovil 519	12.41	Taunton
11.14 Salisbury	12.41	BR5 4-6-0	N. Elms 22		Loco to MPD
Loco ex-MPD		M7 0-4-4T	Yeovil 517	13.05	Junction
13.15 Junction	13.19	M7 0-4-4T	Yeovil 517	13.25	Junction
Loco ex-MPD		S15 4-6-0	Sal 497	13.35	Exeter
13.40 Junction	13.44	M7 0-4-4T	Yeovil 517	13.56	Junction
13.23 Langport	13.51	45xx 2-6-2T	Taunton 60	13.52	Pen Mill
14.10 Junction	14.14	M7 0-4-4T	Yeovil 517	14.23	Junction
14.32 Pen Mill	14.34	5700 0-6-0PT	Yeovil 518	14.36	Taunton
14.36 Junction	14.40	M7 0-4-4T	Yeovil 517	14.56	Junction
14.00 Taunton	15.13	5700 0-6-0PT	Yeovil 519	05.16	Pen Mill
15.10 Junction	15.14	M7 0-4-4T	Yeovil 517	15.30	Junction
13.08 Exeter	15.40	S15 4-6-0	Ex-Jcn 545	Forms -16.06 to Jcn	
15.51 Junction	15.55	M7 0-4-4T	Yeovil 517		
16.00 Pen Mill	16.02	45xx 2-6-2T	Taunton 60	16.04	Taunton
		S15 4-6-0	Ex-Jcn 545	16.06	Junction
		M7 0-4-4T	Yeovil 517	16.19	Junction
16.33 Junction	16.37	M7 0-4-4T	Yeovil 51	16.56	Junction
17.08 Junction	17.12	M7 0-4-4T	Yeovil 517	17.18	Junction
16.12 Taunton	17.25	5700 0-6-0PT	Yeovil 518	17.33	Pen Mill
17.28 Junction	17.32	M7 0-4-4T	Yeovil 517	17.50	Junction
17.45 Pen Mill	17.47	5700 0-6-0PT	Yeovil 519	17.50	Taunton
18.05 Junction	18.09	M7 0-4-4T	Yeovil 517	18.20 Junction	
18.33 Junction	18.37	M7 0-4-4T	Yeovil 517		
17.50 Taunton	18.54	45xx 2-6-2T	Taunton 59	18.56	Pen Mill
		M7 0-4-4T	Yeovil 517	19.01	Junction
19.14 Junction	19.18	M7 0-4-4T	Yeovil 517		Loco to MPD
16.48 Basingstoke	19.40	S15 4-6-0	Sal 500		Loco to MPD
19.45 Pen Mill	19.47	45xx 2-6-2T	Taunton 59	19.50	Taunton
Loco ex-MPD		M7 0-4-4T	Yeovil 517	20.10	Junction

Train	Arr	Loco	Diagram	Dep	Destination
20.25 Junction	20.29	M7 0-4-4T	Yeovil 517	20.40	Junction
20.56 Junction	21.00	M7 0-4-4T	Yeovil 517	21.05	Junction
20.20 Taunton	21.11	5700 0-6-0PT	Yeovil 519	21.13	Pen Mill
21.20 Junction	21.24	M7 0-4-4T	Yeovil 517	21.29	Junction
21.40 Junction	21.44	M7 0-4-4T	Yeovil 517	21.50	Pen Mill
21.58 Pen Mill	22.00	M7 0-4-0T	Yeovil 517		Loco to MPD
20.41 Salisbury	22.09	WC 4-6-2	Sal 503		Loco to MPD
18.54 Waterloo	23.24	WC 4-6-2	Ex-Jcn 537		Loco to MPD

Above: 'N15' 4-6-0 No S747 *Elaine* stands on shed at Yeovil Town on 15 September 1948. This locomotive carried the temporary 'S' prefix from 24 January 1948 until renumbered in the 30,000 series on 12 May 1950. The locomotive, dating from July 1922, was withdrawn in October 1956 having attained a total of 1,296,927 miles. The name *Elaine* later passed to BR Class 5MT 4-6-0 No 73119, a locomotive of lesser pedigree and performance. *A. E. West*

Right: A Maunsell 'U' class 2-6-0 has been lifted by the gantry at Yeovil Town for attention by the shed fitters on 13 August 1946. Class B4 0-4-0T No 94 stands in the foreground. The 'B4s' were not allocated to Yeovil and as this particular locomotive was based at Plymouth Friary and is pointing 'down' the road (smokebox towards Exeter) she is probably returning to her home depot from overhaul at Eastleigh. The coaches to the left form the breakdown set carrying jacks and packing equipment for minor derailments. Breakdown cranes were also kept at Salisbury and Exmouth Junction. *A. E. West*

Yeovil Town Locomotive Shed

Nestled between the station and the lee of Summerhouse Hill, the brick-built shed with a slate-hung pitch roof had three roads. With its aroma of smoke and steam mingling with the smell from the nearby gasworks, it was one of the busiest medium-sized depots on the LSWR, at one time being responsible for the sub sheds at Templecombe and Chard Town. Staff numbers in 1909 comprised a foreman, two clerks, 30 drivers, 27 firemen, 21 engine cleaners, one cleaner chargehand, three fitters, three fitters' lads and one boilersmith, as well as other ancillary workers. Twenty-three locomotives were stationed at the shed and breakdown vans containing ramps and lifting jacks etc were also stabled there. Inside the building all three roads had pits, as did Nos 1 and 2 roads, the table and the coal roads outside. A water tank was provided inside the building which ran across the roof of the shed stores, water being pumped by a stationary engine from a well sunk near Dodham Brook.

A 43ft turntable was installed, but over the passage of time it was found to be rather small for the increasing size of new locomotives and was incapable of enlargement due to its location. The new turntable at Yeovil Junction, dating from the 1907-08 remodelling, was by then in use, and an LSWR minute of 1917 describes the shed turntable as 'out of repair and no longer required'. It was removed shortly afterwards and a siding extended from the former head shunt of the turntable to run alongside the shed was known as the table road until the closure of the depot. Coaling in my day was accomplished by a steam crane which appeared at the shed after the World War 2. Until then it had been done by hand from the coaling stage, for which a shelter was provided in 1920 at a cost of £176 plus £6 10s for lighting. My early engine cleaning days were spent in a brick-built cabin standing alongside the main coaling road, this building having replaced an old van body which had stood there since 1917. The junior engine cleaners at Yeovil were always known as 'nippers' or 'nips', and this name would stick. I well remember a father and son who were driver and fireman respectively, and the father, when looking for his son one day, asked 'have you seen our nip?' His son was around 30 years of age!

Above: Yeovil Loco, as I will always remember it. In this view taken from inside the shed on 17 August 1963, No 5563 stands on the No 2 shed road alongside Maunsell 'S15' 4-6-0 No 30828. The bunker of No 4593 is seen to the left, while a Bulleid Light Pacific can be glimpsed standing outside on the table road. *Ben Ashworth*

In addition to the fitters and boilersmith, two storemen and a boilerwasher were employed in the shed as well as three sets of coalmen on the early, late and night shifts. The stores were a haven of warmth on a cold day with the tanks of lubricating oil being warmed by gas jets to keep the contents supple. It was the fireman's job, while preparing an engine, to collect the supplies of lubricating and engine oil plus paraffin for the lamps. Measures were carefully issued by the storeman who then entered the engine number in the ledger. Split-cane corks, 'rustys' (honeycombed cloths issued from Eastleigh), firing shovels, headlamps and all the equipment used on a steam locomotive was also issued there. In the 1960s, the complement of drivers, passed firemen, firemen and engine cleaners totalled 50 men; with the shed staff, coalmen, shed labourers and clerical staff added, this brought a total of approximately 100 employed at the depot. The depot was swelled with the addition of the Pen Mill loco crews and 10 tank engines when the ex-GWR shed closed in 1959. An additional water tank was erected at the rear of the building, and the old LSWR tank abandoned. An extra water crane was also erected and the drivers' cabin extended. A certain young fireman I know of used to swim in the water tank on the odd occasion!

With the arrival of the Western men, came the extra work to be absorbed with our Salisbury-Exeter Southern main line duties, including branch line services to Taunton, banking at Evershot and Brewham, shunting at Durston and Hendford, freight trains to Westbury, and excursion trains to Weymouth; all of which made our depot a very busy and interesting place to work. Incidentally the depot was always known as 'Yeovil Loco' and never Yeovil shed or motive power depot. I remember my days spent working there with great fondness, although at times, if we were having a rough trip on a very wet, early morning freight climbing Honiton bank, then that was the opposite side of the coin. I count myself very fortunate indeed to have worked on some of the finest steam power ever to have been seen in this country.

A depot the size of Yeovil Town had to cater for many locomotive diagrams. These were very complex-and would be operated by two, three or more sets of men throughout a 24-hour period. Yeovil footplate crews' weekday duties for the summer of 1963 comprised Yeovil Duties Nos 1 (P & D), 512,

Above: Drummond Class C8 4-4-0 No 294 stands under the hoist on the No 1 road inside Yeovil Town Loco on 12 June 1926 shortly before being reallocated to Fratton. Although fitted with a Urie smokebox door, the locomotive still retains its splasher sandboxes. No 294 worked a special train from Windsor to Dover carrying the Grand Duke and Duchess of Serge upon their return to Russia in November 1899. The train was piloted on the LC&DR by 2-4-0 No 49 *Zephyr*. The gas lamp on the wall was still extant (but not working) when I entered the railway service in the 1950s. *H. C. Casserley*

513, 514, 516, 517, 518, 519, 520, 521. Salisbury Duties Nos 461, 471, 478, 479, 480, 497, 500, 501, 503, 505, 506. Exmouth Junction Duties Nos 544, 545, 546, 547, 548, 550, 551, 561, 584, 607. The branch services to Taunton were worked under Yeovil Duty No 518, and Taunton Duties Nos 31 and 33; the Taunton-Yeovil freight (6.20am ex-Taunton) came under Plymouth Duty No 834 (Taunton men). When the Class 22 diesels started working the Taunton-Yeovil services they were worked by Taunton crews under Newton Abbot Duty No 46.

Two 2-2-2s constructed by Stothert & Slaughter for the Italian-Austrian Railway, *Ombrone* and *Bizenzia*, were delivered to Nine Elms for running in trials with the approval of the LSWR. Having successfully completed their trials and awaiting transhipment from Southampton Docks, the news came that a shortage of funds had delayed the opening of the Italian-Austrian line and, as a consequence, both locomotives were offered to the LSWR and accepted. Renamed *Gem* and *Ruby*, they were placed at work on 12 December 1847. From September 1859, *Ruby* was transferred to Stoford (Yeovil Junction) for ballasting the line to the temporary terminus at Hendford. *Gem* was also transferred to Yeovil at a later date from where withdrawal occurred in May 1862. Nineteen engines were transferred to Exeter for the opening of the line in July 1860, including Rothwell 2-2-2s *Fireball, Harpy, Pegasus* and *Wildfire*; Fairbairn 2-2-2s *Acheron, Achilles* and *Actaeon*; and 'Bison' class six-coupled goods *Lioness, Leopard* and *Panther*. For the opening of the line, the directors and dignitaries of the LSWR left Waterloo at 8am on 18 July 1860 in a 20-carriage special headed by *Britannia* and *Vulcan*, both 'Etna' class singles, and *Montrose* of the 'Canute' class. They were joined en route by officials of the Salisbury & Yeovil Railway and reached Stoford (Yeovil Junction) just after noon, having stopped at each station on the line to Exeter to receive and answer congratulatory addresses from local dignitaries. There is an interesting account regarding the well-known photograph of the locomotive *Frome*, a Christie, Adams & Hill 2-2-2 taken at Yeovil in August 1862. Driver Thomas Hicks had commissioned a local photographer to take the picture of himself and his fireman on the footplate on a sunny Sunday morning. Unfortunately the powers that be summoned the driver to Nine Elms to explain 'his

27

Above: 'U' class 2-6-0 No 31794 stands on the No 2 road inside Yeovil Town shed on 10 July 1956. One of the 'M7' branch tanks is being cleaned alongside. Steam drifts hazily around the shed where I and countless engine cleaners have polished away at grimy engines before our entry into the footplate grades. *R. C. Riley*

illegal use of the company's engine *Frome*, coke, oil and water on Sunday last'. A severe reprimand was issued, and the photographs confiscated and destroyed (but not the negative). The locomotive also hauled the eight-coach opening train from Gillingham to Sherborne on 7 May 1860 from which the directors received a rapturous welcome.

Tiger, a 'Bison' class 0-6-0 allocated to Yeovil, was involved in a derailment on the main line within a short time of the opening. The locomotive was required to work a cattle special from Exeter to Salisbury on 29 July 1860 and, to save line occupation, was double-heading with *Colne*, a 6ft 2-4-0 of the 'Tweed' class, on the 3.30pm Yeovil-Exeter stopping train consisting of six carriages and two vans. The tender of the leading engine suddenly left the rails between Honiton and Feniton causing *Tiger* and the train to crash down the embankment. No serious injuries occurred, but the leading axle of *Tiger* was broken and the firebox punctured. The subsequent Board of Trade inquiry determined the cause of the accident as excessively fast running on a line that had been open for only 10 days. *Colne* is also recorded as working the first Salisbury-Yeovil Junction cheap evening excursion on 22 September 1860.

Marmion of the 'Canute' class, also a Yeovil engine, was in trouble on 20 November 1860 while hauling the 6.5pm Salisbury-Yeovil passenger train consisting of four carriages and a brake van weighing 57½ tons. When about 4¾ miles from Gillingham, and travelling at a speed of approximately 55 to 60 mph, the locomotive left the rails and came to rest overhanging an underbridge. No one was seriously injured and damage to the locomotive was slight; yet again the blame for the accident was put down to excessive speed on a line recently opened to traffic.

Napoleon and *Montrose*, two further locomotives of the same class, were also based at Yeovil in 1860. Three of the Gooch 'Mazeppa' class singles, 2-2-2s Nos 53 *Mazeppa*, 58 *Sultan* and 61 *Snake,* came to Yeovil when the line from Salisbury opened. No 173 *Nymph* of the 'Undine' class arrived in 1863, and Nos 172 *Zephyr* and 175 *Hebe* of the same class were there until the end of their days in 1886. 'Falcon' class *Argus* went to Yeovil new in 1864, while by mid-1873 Nos 25 *Reindeer* and 91 *Spitfire* of the 'Volcano' class had been allocated, working the daily Yeovil-Torrington service which had started in 1872. *Spitfire* is shown to have been travelling daily to Exeter Queen Street with the 7.15am semi-fast, returning to

Above: Nos 5563, 4507 and 31864 share the table road at Yeovil Town shed on 6 October 1962. The large water tank at the rear of the shed was erected in 1959 to cope with the extra locomotives from the ex-GWR shed at Pen Mill which closed in the same year. *R. C. Riley*

Salisbury with 12.55pm up and returning home with the 4.30pm stopper. The shed foreman at Yeovil was ordered on Thursday, 2 April 1874 to ensure that this locomotive, or No 84 *Styx* of the 'Falcon' class (allocated 1874), were regularly rostered for this duty. *Reindeer* was rostered on alternate days to work the 6.15am semi-fast to Waterloo, returning as far as Salisbury with the crack 2.10pm from London. An 0-6-0 goods locomotive of the 'Lion' class, No 16 *Salisbury* (renamed *Stonehenge* in August 1877), arrived new in 1872 and was usually rostered to work the 9.40am semi-fast to Exeter, returning with the 4.20pm.

One of the long-boilered George England 2-4-0 Engineers Department locomotives, No 7 *Hesketh*, was withdrawn in August 1878 after a 'pitch in' at Yeovil following an embankment slip during heavy rain. This engine was replaced by another George England 2-4-0, No 14A, purchased second-hand from the Somerset & Dorset Railway and subsequently becoming No 7 *Fowler*. Single-framed Beyer Peacock goods 0-6-0s Nos 339 and 340 were at Yeovil in 1876, while No 0229 spent her final days there in 1910. Double-framed 0-6-0 Beyer Peacock goods Nos 221, 225, 242 and 286 were allocated in March 1878, with

No 242 condemned from Yeovil in October 1891. Members of the Adams classes were based here, with the '380' class ('Steamroller') 4-4-0s Nos 387, 388 and 389 allocated in the late 1890s, and Nos 0386, 0387, 0388 and 0389 in December 1902. Class 460 4-4-0 No 476 arrived new in 1884, while No 0469 of the same class was regularly seen in 1912 working the heavy 7.20pm up milk to Waterloo and returning double-headed with the 2.40am goods from Nine Elms. Class T3 4-4-0 No 571 arrived when new in 1893, being joined by No 575 the following year, and worked an early morning express to the capital, returning with vans. Members of the sleek and handsome 'X6' class 4-4-0s were at Yeovil in the summer of 1910 with Nos 664 and 666, working in turn, handling the 8.41am Yeovil Junction-Waterloo slow and returning with the 2.25pm Wimbledon-Templecombe milk empties which invariably conveyed 30 vans. Class X2 4-4-0s Nos 577 and 585 are recorded as working milk trains from Town station in August 1918. 'A12' 'Jubilee' class 0-4-2s Nos 551 and 553 went new to Yeovil in 1899, sharing a duty roster that took them to Clapham Junction with a daily van train for over 10 years, returning early the following morning via Eastleigh with a goods. No 638 was also at Yeovil, and in May 1932 Nos 612 and 616 were also allocated. 'G6' class 0-6-0 tank locomotives Nos 140 and 266 were there in 1901, one being sub-shedded at Templecombe, with Nos 275, 276 in June 1929, and Nos 238, 276 and 351 in August 1939.

The famous Drummond 'T9' class 4-4-0s were first allocated to Town shed in June 1902 with Nos 286, 287 and 288 arriving new, 713 and 714 appearing in October 1903, Nos 116, 117 and 300 in 1911, with 113, 716, 721 and 728 in June 1932, Nos 310, 702, 710, 712, 714, and 716 in October 1947, and No 30706 in June 1955. 'K10' class ('Small Hopper') 4-4-0 No 390 appeared in 1904, with Nos 143, 145, 340 and 387 in 1925, Nos 143, 145, 152, 340 and 344 in 1939, with No 389 in 1950 being joined by 384 in 1951. An interesting duty commenced in May 1942 when Exmouth Junction 'T9s' (usually Nos 709, 721 or 724) commenced working between Exeter and Yeovil via the GWR route between Taunton and Martock as part of interchange duties to make more economic use of engines and crews, the GWR balancing duty on a mileage basis being from Exeter Central to Barnstaple and Ilfracombe using '43xx' 2-6-0s. Class L11 ('Large Hopper') 4-4-0 No 165 was at Yeovil in June 1908, with Nos 134, 163 and 436 in 1939, and 134, 163 and 412 in mid-1948. 'S11' class 4-4-0 No 397, after being fitted with superheating at Eastleigh, was transferred to Yeovil to work the London milk vans; No 402 went to Town shed in 1926 to work a daily Portsmouth via Salisbury return duty, and was joined by 397 and 398 in March 1932. However, their stay was brief and they were moved away in mid-1932, being replaced by 'U' class Nos 790-2, (and subsequently 1793-5), but they returned in September 1934 to be joined later by No 404. 'C8' class 4-4-0s Nos 290-4 were transferred from Basingstoke to Yeovil in 1926, only to be sent away by June of the same year. No 295 was at Yeovil in 1932, principally to work an early morning Exeter stopper, returning with vans. However, it was withdrawn in 1935, having lain derelict at Yeovil for a year before withdrawal. Bulleid Pacifics, 'King Arthurs', 'S15s' and BR Standards were common visitors in BR days, although none of the larger classes was ever allocated to Yeovil. The shed yard at weekends was always full of 'foreign' locomotives laying over between rostered duties. The shed code 72C, which had been in force since 1950, changed with the Western Region take-over in December 1962, but the new code 83E was not given official status until September 1963. The shed closed to steam in June 1965 but remained open as a stabling point only, the locomotives being dispersed to Templecombe 83G, Worcester 85A and Oxford 81F.

One of the main features of Yeovil Town throughout the years was the intensive shuttle service worked between there and Yeovil Junction, meeting main line stopping trains. The first locomotive known to have been used was No 108 *Ruby*, on ballasting work at Yeovil and being pressed into service with the opening of the Exeter line on 18 July 1860 on the

services between Yeovil Junction and the temporary station at Hendford. Built by the Bristol firm of Stothert & Slaughter, *Ruby* was a 2-2-2 Gooch-type tender locomotive weighing 26 tons with driving wheels of 6ft 6in, a wheelbase of 13ft 6in and a boiler pressure of 90lb. The locomotive's career was brought to an end on 17 December 1864 when someone at Yeovil lit her firebox with an empty boiler. It was subsequently sent to Nine Elms for breaking up. *Gem*, a sister locomotive, also worked at Yeovil but was withdrawn in May 1882. Two Beattie 2-2-2 well tanks, *Vizier* and *Transit*, worked on the branch until being withdrawn in 1871 and 1872 respectively. Both engines had previously been based at Nine Elms for working Richmond services. *Cossack*, a 'Sussex' class 2-2-2WT was on the branch until 1877; a sister engine, *Princess*, was transferred to Gillingham (Dorset) in 1859 for the opening of the line and worked the shuttle services to and from Salisbury until the section to Templecombe was opened on 7 May 1860. 'Nile' class 2-4-0WT *Cressey*, previously employed on the Waterloo suburban services, was sent to Yeovil for the branch services, local shunting and Templecombe services, until withdrawn in August 1882. Two standard well tanks of the 2-4-0 '329' class, Nos 44 and 220, were in service on the branch in 1890 — No 44 was previously allocated to Twickenham in 1878, with No 220 working from Kingston in the same year.

In the summer of 1903, ex-LBSCR Stroudley 'Terrier' 0-6-0T No 735 (formerly No 668 *Clapham*), one of two of the class purchased by the LSWR, was at Yeovil and in use on the shuttle services for a short time before departing for trials on the Lyme Regis branch in early August of the same year, ahead of the official opening later in the same month. The other LSWR Stroudley 'Terrier', No 734 (No 646 *Newington*), arrived at Yeovil in 1909 for the Town-Junction service before being transferred to Fratton for the Lee-on-Solent line and other duties, thence being hired to the Freshwater, Yarmouth & Newport Railway on the Isle of Wight in 1913 where it was renumbered No 1. The locomotive was purchased by the company in 1915 and was subsequently renumbered No 2 in 1919. After the Grouping, it was painted Maunsell green and renumbered W2 in March 1924, receiving the nameplate *Freshwater* in October 1928. Further renumbering came in 1932 to W8. After returning to service on the mainland for the Hayling Island branch and being displayed outside a public house for a number of years after withdrawal, the locomotive is to be found today working services out of Haven Street station on the Isle of Wight.

The motor train services at Yeovil began in July 1915 using the Drummond cable and pulley system with Adams 'O2' class 0-4-4Ts Nos 216-8. The locomotives were equipped with cables leading across the cab roof and bunker to the trailer cars for the duplicate controls in the trailer driving ends; unfortunately, due to some serious incidents with this

Above: Sandwiched between No 34056 *Croydon* and No 31637 at Yeovil Town on 13 August 1963 is the loco stores van, No DS44385, an ex-LSWR 10-ton covered goods van. Branded as an MP (Motive Power) Stores Van, this vehicle conveyed materials as ordered by the depot storemen between the works at Eastleigh and Yeovil Loco. *A. E. West*

equipment on the Bournemouth West-Wimborne-Brockenhurst services, it was decided in 1929 to convert 31 'M7' and four 'O2' locomotives to the air control system using air pumps and piping from withdrawn LBSCR locomotives at Brighton Works. In the interim, a number of 'D1' class 0-4-2 tanks were fitted with this system and, with their motor sets, were sent to the Western Section — Nos 2299, 2616 and 2273 appearing at Yeovil in the 1930s. Adams 'O2' tanks Nos 187 and 207, converted to the air control system, were allocated in September 1939 to replace 'D1s' Nos 2299 and 2614, but in March 1940 were ousted by 'M7s' Nos 58 and 129. No 129 had previously been allocated to Guildford in 1930 and Fratton in 1937; subsequently renumbered 30129 under Nationalisation, she remained at Yeovil for over 20 years. No 30131 was allocated from Bournemouth in 1951, replacing No 30058, while others, including Nos 30034 and 30036, were at Yeovil at various times between 1950 and 1960. The auto sets varied at times, especially when being overhauled at Eastleigh. One favourite was set No 373 dating from 1914 and fitted with iron gates instead of doors. No 2622, the surviving gated control trailer, was withdrawn in

October 1960. Other two-coach sets, comprising a brake third and brake composite, were originally built in the 1920s and converted for push-and-pull operation in 1949.

All was well at Yeovil until that dreadful day when the Western Region gained control of the lines west of Salisbury, and in March 1963, our Drummond tanks were taken away and replaced by ex-GWR '6400' class 0-6-0PTs, with '1400' class 0-4-2T No 1451 arriving in October 1963 — a poor substitute in my view — but then I was a Southern man, and I will comment no further. The final chapter in the branch services came on 4 January 1965 when the steam services were replaced by diesel railbuses that had been made available by the closure of the Kemble-Cirencester branch. The branch shuttle services ceased running from Yeovil Town on 2 October 1966 and from this date ran between Pen Mill station and the junction; incidentally this was the first regular booked passenger service to use the former wartime connection at Yeovil South Junction since its opening in October 1943. The shuttle services from Pen Mill did not last long, being replaced on 4 May 1968 by a bus.

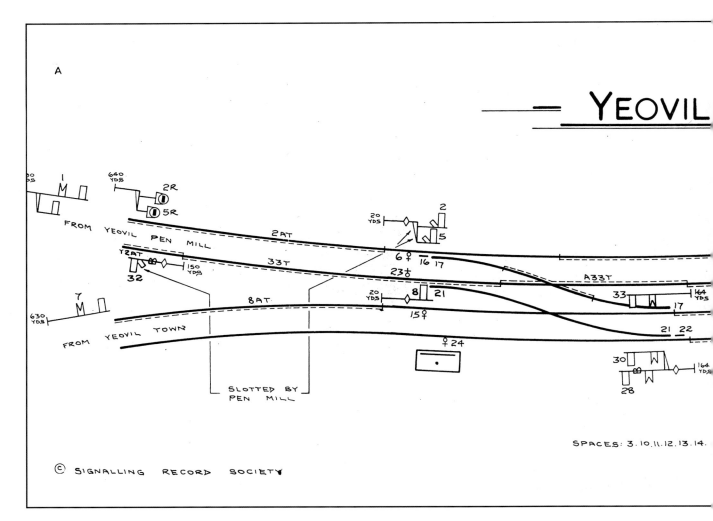

Yeovil Junction

Yeovil Junction and the line to Exeter opened with three trains daily on 19 July 1860, just five weeks after the opening of the Salisbury & Yeovil line to Hendford, the full service starting on 1 August with goods traffic introduced in September the same year. The Act for the Exeter Extension had been obtained on 21 July 1856, the authorised line being from the Salisbury & Yeovil at Bradford Abbas to Exeter Queen Street, with a junction near Bradford Abbas throwing a spur line to the S&Y nearer Yeovil to complete a triangle. The LSWR timetable for August 1860 shows certain trains, such as the 7.30am from Exeter, stopping at Yeovil Junction before running via the spur to the temporary station at Hendford, thence reversing to travel direct to Sherborne via Yeovil Upper Junction and Bradford Abbas Junction. Other services, like the 1.5pm from Exeter, travelled direct to Sherborne and Salisbury and were fed at Yeovil Junction by a connecting service. The section between Bradford Abbas and Yeovil Junction was doubled in 1861 and between Yeovil Junction and Yeovil Town in 1864, the whole route between Salisbury and Exeter being doubled by 1 July 1870. The original S&Y easterly curve through Bradford Abbas was closed on 1 January the same year, although the bridge which carried the route over the GWR remained in place

until 1937. From that date, trains to and from Salisbury via Yeovil Town had to reverse at Yeovil Junction. In order to improve efficiency at the congested Hendford goods depot, the GWR constructed a broad gauge, single-track goods branch, known as the Clifton Maybank Siding, branching off the Weymouth line and terminating in exchange sidings and a transhipment shed on the down side at Yeovil Junction. This opened on 13 June 1864 and closed on 7 June 1937.

The Salisbury & Yeovil Railway, while financially successful, lacked attention to passenger amenities at its stations, the LSWR having to insist that the S&Y provide shelter at its smaller stations and lengthen platforms to 250ft to prevent difficulties and delays. In 1871, with longer trains being commonplace, it only agreed to extend them another 50ft on finding that the LSWR stations were treated likewise! The line between Salisbury and Exeter soon became one of the busiest and most profitable for the LSWR which finally gained control of the Salisbury & Yeovil in 1878, thereby making a handsome profit for the S&Y shareholders.

The Drummond '700' class 0-6-0 goods engines suffered from teething troubles when new and, amongst several defects, it was found that the regulators suffered from the unwelcome effect of jamming open with the result that on 13 July 1897 No 690 crashed into the six-coach 6.40am Yeovil

TO YETMINSTER

TO YEOVIL JCN 'A'

693
YDS

693
YDS

4

9

25

J.P.M.   18. 7. 8

DOWN
UP   MAIN
DOWN
UP   S.R.

Junction-Yeovil Town while backing on. Ten passengers were injured, and damage to the coaches included lamps and windows broken, and screw couplings twisted. Inspection of the locomotive found that the regulator assembly deviated from the drawings, and this was rectified by the works at Nine Elms at the expense of Dübs & Co, the manufacturers. Another accident occurred on 4 July 1915 when the 10.25pm freight from Exeter to Salisbury broke away at Sutton Bingham. Before the leading portion could be diverted into the loop at Yeovil Junction, the runaway vehicles collided with the front portion approximately 200 yards from the West box, wrecking 22 wagons and demolishing part of the box.

In November 1905, the Engineering Committee of the LSWR authorised expenditure of £37,030 to strengthen the permanent way and bridges along the Salisbury-Exeter route in anticipation of the arrival of the Drummond 'F13' class 4-6-0s. It was also now realised that approaching competition with the GWR cut-off routes to the West would place the LSWR in an unfavourable position. The old station at Yeovil Junction with its two narrow platforms, cramped wooden buildings, a single line for the branch trains in the middle, which had to cross the main line every time they arrived at and departed from the station, and sharp curves with severe 20mph speed restrictions, was fast becoming a bottleneck for the

expanding services to and from the West. Consequently the whole layout was drastically remodelled during 1907-9 at a cost to the South Western of £47,400; two new platforms were provided with four main running roads through the station with the branch trains now running into a bay alongside the up platform. The layout was completely resignalled with a new 60-lever East cabin located in the vee created by the Yeovil Town branch and the up main line, and a new frame installed in the existing West box. Several acres of land were purchased and part of a sandstone cliff cut away to extend the siding accommodation. Major Pringle inspected the completed layout on 3 March 1909. From 19 July 1926, when the 11am from Waterloo was named the 'Atlantic Coast Express', the Sidmouth and Exmouth coaches were detached at Yeovil Junction, and normally there was a corridor brake for Sidmouth and a similar vehicle for Exmouth. In the reverse direction, the Sidmouth and Exmouth coaches were attached to the up 'ACE' at Yeovil Junction. However, from the summer of 1928, the down 'ACE' stopped at Sidmouth Junction instead of Yeovil Junction.

The turntable, which has surprisingly survived since steam days, being used occasionally by the permanent way department, is still in use. It has been restored and renovated by the volunteers of the South West Main Line Steam Company, a small but highly

33

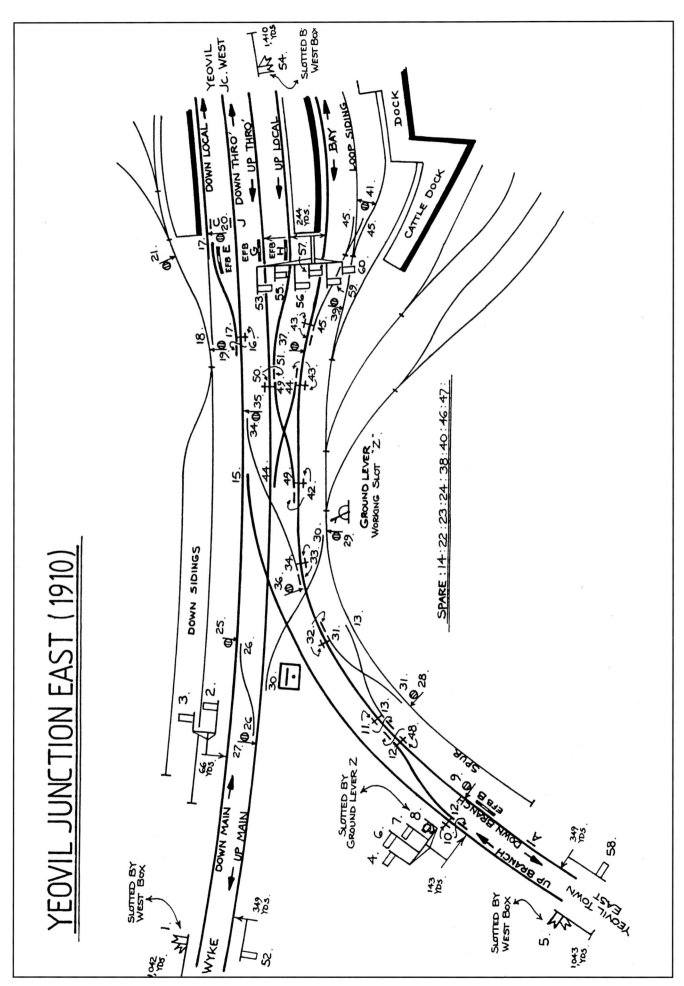

YEOVIL JUNCTION EAST (1910)

SPARE : 14 : 22 : 23 : 24 : 38 : 40 : 46 : 47

Fresh off Yeovil Town Loco and with plenty of steam to spare for the gradients ahead, 'S15' No 30828 stands alongside the down platform at Yeovil Junction on 17 August 1963 with the 1.35pm Yeovil Town to Seaton. *Ben Ashworth*

Above: 'M7' No 30131 propels ex-LSWR motor set No 373 into Yeovil Junction with the shuttle from Yeovil Town on 22 August 1959. *A. E. West*

Below: The shuttle service to Yeovil Town with 'M7' 0-4-4T No 30131 and 'gate' set No 373 departs from the branch platform at Yeovil Junction on 23 August 1959. Not only is the branch bay to Yeovil Town starter in the off position (far right) on the gantry, but a train is also signalled on the up through (far left). *A. E. West*

Above: Maunsell 'U' class 2-6-0 No 1792 from Yeovil Town shed stands at Yeovil Junction with the 11.16am Exeter to Salisbury on 21 May 1935. A private owner wagon belonging to the well-known local coal merchants of Messrs Bradfords can be seen in the background. *H. C. Casserley*

Below: Driver Maurice Gerrard and fireman Ron Saunders — both from Yeovil Town Loco — refill the tanks of auto-fitted 5400 class 0-6-0PT No 5410 at Yeovil Junction on 17 August 1963 before departing with the 12.45pm to Yeovil Town. *Ben Ashworth*

efficient group of enthusiasts who have relaid track to allow their 0-4-0 saddle tank *Pectin* to give visitors brake van rides on open days from the former down platform. An English Electric Class 20 diesel, No D8188 *River Yeo*, resplendent in black livery, is also based here. On-site watering facilities for visiting steam locomotives have been restored, and a railway video and bookshop provided in the former refreshment rooms on the down platform (all South West Trains services use the former up platform), but the crowning achievement at the Yeovil Railway Centre must be the completion of a superb new engine shed located behind the down platform on the site of the old GWR Clifton Maybank goods station. I never imagined when I left the footplate in 1964 that I would ever see an engine shed built in the area again, especially when our old shed at Yeovil Town was destroyed along with the station. More extensive details regarding Yeovil Junction and the Clifton Maybank line, plus the other main line junction stations and the Fovant Military Railway, can be found in the companion volume to this book — *The Salisbury to Exeter Line* — by Derek Phillips and George Pryer (OPC).

Footplate Memories

The shuttle services in my firing days at Yeovil Town were handled almost exclusively by 'M7' Drummond and 'O2' Adams 0-4-4 tanks during the mid-1950s, and until 1963. Fitted for push-and-pull working, the locomotives were known to the footplatemen as the branch tanks or motor tanks, and *never* called the 'bunk' or 'bucket' as was done by the local train spotting fraternity. The services were usually crewed by men not allowed on the main line for medical or disability reasons but, as often happened, when the regular men were on holiday or sick etc then it was down to the men in the spare link to provide cover.

Let us now travel back through the years to my old shed at Town station on a summer Saturday in 1961.

The shed yard is packed full of engines as we walk over the tracks heading towards the shed office to book on; the air is heavy with the aroma of coal smoke and steam, this being compounded with the pungent smell from the nearby gasworks. Engines are being prepared and disposed; hot clinker cascades down into the pits from a Bulleid Pacific, and a nearby 'U' class 2-6-0 lifts her safety valves with a tremendous roar as she is prepared. Other locomotives

are being prepared for work or disposed as they return from duties on the main line. The steam crane rattles into life, hoisting heavy tubs of coal and depositing the contents upon the tender of a 'King Arthur' standing on the No 1 shed road, then drops the empty iron tub with a thump on to the ground alongside a coal truck in which the coalmen are feverishly working at their task refilling the tubs, while another locomotive trundles down the coal road to replenish its tender. Three roads, known as Nos 1, 2 and 3 (from left to right), serve the shed. The long road to the side of the shed by Dodham Brook is always known as the table road, although the turntable has been removed long ago. After booking on at the shed office at 6.50am (the late shift crew would book on at 2.20pm and relieve us at 2.35pm) and picking my kit box from the locker room, it's now time to enter the drivers' cabin situated alongside the shed office building. The air is heavy with tobacco smoke from numerous cigarettes and pipes, and there is the usual exchange of banter from the men seated at the long table upon which are placed the day's newspapers — courtesy of the early morning paper train! — dirty tea mugs, the remains of someone's sandwiches and fag ends. It is within the enginemen's cabins that all kinds of debate will, at times, be raging. I have known it to be the same at all steam sheds at which I have had the pleasure to work, including Salisbury and Exmouth Junction, and it is fair to state that I have never worked, before or since, with such a superb group of people as the steam footplatemen, a band of brothers unlike any other. The atmosphere in our small shed at Yeovil is quite parochial, and if anyone is in trouble of any kind then we will do our best to try to help. We also have our own football team (Railway United) and a skittle team which plays at the Alexandra Hotel (known as the Alex), situated opposite the station entrance. Happily, it will still be trading in the year 2000!

Now it's time to join our engine and, as we leave the cabin, there she is — 'M7' No 30131 in her coat of lined black standing out in the shed yard. She is already prepared and the safety valves are lifting with that wonderful sizzle that the Drummond locomotives have, completely different from others with safety valves that lift with a shattering roar. The shed turners, in order to save space in the shed yard, have already placed the locomotive on to the two-coach set which is stabled on the road behind Platform 3. With a bit of luck we will be already coupled up but if not, the shunter will soon appear. Failing that, then it will

Above left: 'West Country' class Pacific No 34006 *Bude* arrives at Yeovil Junction with a train of empty ballast hoppers for Meldon Quarry on 18 February 1965. A BR Standard 2-6-4T stands on the down sidings. *A. E. West*

Left: No 5410 has shut off steam and drifts into Yeovil Junction while propelling its two-car set from Yeovil Town. The SR main line in the foreground crosses over the WR Castle Cary-Weymouth line. Weeds and grass cover the formation of the GWR Clifton Maybank goods line, which lay alongside the GWR main line before burrowing under the Salisbury-Exeter line and rising to the junction station. *Ben Ashworth*

Above: A general view of Yeovil Town station on 20 September 1964. The Southern lines are in the foreground, with the ex-GWR line to Pen Mill on the left. The covered bay on the far left was used for loading parcel vans. The station building was destroyed by the town council — reflecting a society bereft of foresight — and turned into a car park. Plans for the year 2000 involve the council selling off a large part of the car park to a private developer — the area is to be used for a multi-screen theatre and 10-pin bowling complex. I politely refrain from further comment. *A. E. West*

be down to me. As well as the vacuum and steam heating pipes (winter only) there will be three air hoses to connect, these being marked — 'back pressure', 'main storage' and 'regulator control', plus the electrical connection for the bell system between driver and fireman. The branch tanks at Yeovil are meticulously looked after, both mechanically and in the cleaning department; the night engine cleaners would have polished the engine framing to perfection (I have done it many times while engine cleaning). Normally one engine is kept in steam while the other is being washed out etc — the two favourite 'M7s' being Nos 30129 and 30131. My mate and I climb aboard and put our boxes away; the cab is hot, with a roaring from the firebox as the fire burns away. The smell of hot oil is everywhere; water dances up and down in the gauge glasses, and the boiler front and brass gauges gleam away. My mate opens the large ejector and the vacuum gauge quivers into life and holds at the required 21 inches. With a loud hiss, the

handle is brought down and the vacuum destroyed; now it's time to unscrew the handbrake with a glimpse over the side to check that no other locomotives are moving near us. With a blast on the whistle we pull out from the shed yard with our two green coaches in tow and trundle past the shed starting signal at our booked departure time of 7.5am on Yeovil Duty 517.

We pull up over the crossover and, when the road is given, reverse back into one of the platform roads, usually number 2. The beauty of Town station is that trains departing for Yeovil Junction can leave from any one of the three platforms, giving flexibility of working in busy periods. We coast into the platform for our first trip of the day at 7.22am. However we are not the first passenger train out from Town as the 6.25am Yeovil Town to Ilfracombe and the 6.55am to Salisbury have preceded us. It is traditional on the branch services at Yeovil to have the driving trailer facing towards Yeovil Junction with the locomotive

Above: With smoke drifting lazily from the chimney, 'M7' 0-4-4T No S58 and push and pull set No 659 (ex-SECR) stand within the confines of Yeovil Town station on 15 September 1948. This locomotive carried the 'S' prefix from 13 March 1948 until 15 December 1950. No 58, dating from March 1906, was fitted with the air-operated push-and-pull system in June 1930 and was withdrawn in September 1960. *A. E. West*

propelling bunker first. My mate walks back to the driving trailer, and it's now time to turn the brass wheel and so engage the Westinghouse pump bolted to the fireman's side of the smokebox. The loud panting noise from the air pump echoes around the station; fast at first, then slower as the air pressure builds up. The auto regulator gear is already in place, the vacuum is off, and the reverser in full reverse gear. We are ready to leave; the fire is burning through nicely, the damper is open and there is about three-quarters of a glass of water in the gauge glasses. I keep an eye out on the platform side, ready to slam on the vacuum brake just in case any late comer runs for the train, misses their footing and is unlucky enough to fall under the train. We don't carry a guard as the regulations state that trains between Town and Junction not exceeding three coaches do not need one.

There is a loud ring from the bell; the button on the bell case is pressed to acknowledge the call back to

the driver. We start to move and the auto regulator apparatus comes into life, controlled by my mate in the driving trailer. Now the locomotive barks into life as we leave the station, and with a muffled roar from our exhaust, passes under the footbridge and roadbridge. The station advanced starter and South Junction distant arms are off as we gather speed and rattle over the 62-yard-long viaduct carrying the track over the River Yeo. The footplate starts rocking; the locomotive vibrates with power as every piece of steel seems to come to life while smoke and exhaust steam smother the front end. There is no need to add to the fire on our 1¾-mile journey and the water in the boiler is kept down for two reasons: first, the Drummond tanks do not like the boiler crammed with water, and secondly, it keeps the engine quiet while standing at the stations. At the same time the injectors on the 'M7s' are rather slow and this is where the skill of any particular fireman comes into play. The fire on all Drummonds is kept low and bright, and

41

just before departure, three or four shovels of coal are dropped just under the door, the blast around the firebox upon departure giving us a good head of steam for our short journey. The 'M7s' have a good turn of speed with their 5ft 7in driving wheels and are extremely sure footed — 'strong in the links' is the phrase used by locomen — and our locomotive is no exception as she runs like a well-oiled sewing machine. Built in 1911 at Eastleigh, she was fitted with the air control system in December 1930 and is as good as new. The only problem is that the footplate is a bit cramped and is awkward, especially when swinging clinker shovels around the cab while squaring up the fire on disposal duties but, apart from that, they are good locomotives to work on. We rattle past the wartime signalbox at Yeovil South Junction (which is switched out for most of the time); the Weymouth line appears running alongside, and the sound of our exhaust reverberates along the sandstone cutting as we pass the myriad signals guarding the wartime connection. The Weymouth line disappears from view as our track curves and starts to climb towards the junction; the signalbox and station appear in the distance with our 'M7' still going well. A glance through the round cab windows shows a long trail of exhaust steam beating out from the tall Drummond chimney and drifting back to South Junction; a look at the gauges shows that we have plenty of steam, and the vacuum and air gauges are correct. Now is the time to turn the blower valve on a touch to prevent a blow-back when the regulator is closed; the injector is turned on to top the boiler up, and no sooner is this done when two loud rings on the bell echo around the cab as my mate in the driving trailer shuts the regulator. The gantry appears, with the signal arm raised to give us entry to the branch bay. We trundle and sway over the points and crossings; freight wagons and coaches stand in the sidings, the vacuum gauge pressure lowers as my mate works the brakes. The brakes start to bite on the engine and coaches as we slow rapidly and eventually stop at the platform with a squeal of brakes, the Westinghouse pump still panting away. Time to turn it off, as it will not be needed on the return trip to Town.

We have arrived at our booked time of 7.36am and now pick up passengers detrained from the 6.30am Exeter-Waterloo due in at 7.30am (passengers from Yeovil Town for this train would have been brought out on the 6.55am Yeovil Town-Salisbury) and at 7.42am we depart for Town station to connect with the 6.45am from Taunton due in at 7.48am. We will be on our way back to Yeovil Junction at 7.58am to connect with the 6.40am Exeter-Templecombe. Each day the shuttles make 25 return trips to and from the Junction, also two return trips to Pen Mill, and at certain times we will convey a through coach to the Junction, plus occasionally cattle wagons full of animals from the local Friday market. Incidentally, the cattle are herded down through the streets of the town to the station, much to the consternation of ladies out shopping! Water is taken at Yeovil Junction at various times throughout the day and the locomotive is booked into Yeovil Loco for engine requirements between 12.40 and 12.55pm, to be back in time to work the 1.5pm Town-Junction, and between 7.25 and 8pm to be ready for the 8.10pm to the junction. The last trip of the day will be the 9.50pm to Pen Mill, returning at 9.55pm with the engine booked on shed at 10.10pm at the end of a 15-hour day with a crew change. The longest break in the day will be the 35-minute wait at Town station while awaiting departure with the 10am shuttle to connect with the 7.30am Waterloo-Padstow and the 6.15am Plymouth-Templecombe. The shuttle services achieve 89 miles per day. At one time the shuttle ran an early morning trip to Templecombe and return. On the very rare occasion when the push-and-pull locomotive is unavailable for some reason, then look out, for the fireman has to uncouple and couple up at every trip, which is no joke, especially when time is at a premium. Speaking for myself as a then young fireman, I found the branch work monotonous and at times boring, and was always pleased to return to main line duties.

Above right: Yeovil Town, Sunday, 21 April 1963. No 5416 propels a two-car auto set into the station for the 1.2pm service to Yeovil Junction. A Bulleid Pacific has arrived from Yeovil Junction and locomotives stand in the locomotive yard to the right. *Hugh Ballantyne*

Below right: This is my favourite view of the branch shuttle. 'M7' No 30131 propels 'gated' set No 373 on the 10.18am Yeovil Town-Yeovil Junction past the wartime signalbox at Yeovil South Junction on 2 August 1959. *S. C. Nash*

CHARD BRANCH.

WEEKDAYS ONLY.

(Worked by Great Western Rly. Co.)

SUMMER SERVICE, 18th JULY to 10th SEPTEMBER ONLY.

		To Chard.		—	3m. 00c.	3m. 28c.			From Chard.		—	28c.	3m. 28c.
	Distances.			Chard Junc.	Chard Town.	Chard.		Distances.			Chard.	Chard Town.	Chard Junc.
				dep.	dep.	arr.					dep.	dep.	arr.
a.m.	Pass.	...	...	7 40	...	7 49	a.m.	Pass.	...	...	7 15	...	7 23
„	Pass.	...	A	8 28	...	8 37	„	Pass.	...	...	8 0	...	8 8
„	Empty	...	...	...	9 22	9 24	„	Freight	...	...	9 5	9‡ 7	...
„	Mixed	...	...	10 24	...	10 34	„	Pass.	...	...	9 34	...	9 42
„	Engine	...	...	...	11† 0	11 2	„	Freight	...	...	10 46	10‡48	...
„	Pass.	...	...	11 37	...	11 46	„	Pass.	...	...	11 12	...	11 20
p.m.	Mixed	...	...	12 52	...	1 2	p.m.	Pass.	...	NS	12 7	...	12 15
„	Engine and van	...	...	...	1†50	1 52	„	Pass.	...	SO	12 34	...	12 42
„	Pass.	...	NS	2 42	...	2 51	„	Freight	...	...	1 18	1‡20	...
„	Pass.	...	SO	3 2	...	3 11	„	Pass.	...	NS	2 5	...	2 13
„	Engine	...	NSQ	...	3 23	3 25	„	Pass.	...	SO	2 14	...	2 22
„	Engine	...	SO	3 54	...	4 6	„	Engine	...	NSQ	3 0	3‡ 2	...
„	Pass.	...	B	5 0	...	5 9	„	Freight	...	SO	3 20	3‡22	...
„	Pass.	...	...	5 53	...	6 2	„	Freight	...	SO	...	3 35	3 45
„	Pass.	...	NFS	6 24	...	6 33	„	Pass.	...	SO	4 14	...	4 22
„	Pass.	...	FSO	6 38	...	6 47	„	Pass.	...	NS	4 20	...	4 28
„	Engine	...	NS	7 55	...	8 4	„	Pass.	...	...	5 37	...	5 45
„	Pass.	...	...	9 0	...	9 9	„	Pass.	...	NFS	6 8	...	6 16
							„	Pass.	...	FSO	6 13	...	6 21
							„	Freight	...	NFS	{ 6 43 / ...	6‡45 / 7† 5	7 15
							„	Freight	...	FO	{ 6 55 / ...	6‡57 / 7†14	7 24
							„	Pass.	...	...	8 40	...	8 48

† Depart.   ‡ Arrive.   A—Runs as "Mixed" train when required for the conveyance of live stock only.   B—Horsebox, etc., traffic not to be conveyed by this train.

CHARD BRANCH.

WEEKDAYS ONLY.

(Worked by Great Western Rly. Co.)

WINTER SERVICE, COMMENCING 12th SEPTEMBER.

		To Chard.		—	3m. 00c.	3m. 28c.			From Chard.		—	28c.	3m. 28c.
	Distances.			Chard Junc.	Chard Town.	Chard.		Distances.			Chard	Chard Town.	Chard Junc.
				dep.	dep.	arr.					dep.	dep.	arr.
a.m.	Pass.	...	...	7 40	...	7 49	a.m.	Pass.	...	...	7 15	...	7 23
„	Pass.	...	A	8 28	...	8 37	„	Pass.	...	...	8 0	...	8 8
„	Empty	...	...	...	9 22	9 24	„	Freight	...	...	9 5	9‡ 7	...
„	Mixed	...	...	10 24	...	10 34	„	Pass.	...	...	9 34	...	9 42
„	Engine	...	...	...	11† 5	11 7	„	Freight	...	...	10 46	10‡48	...
„	Pass.	...	...	11 37	...	11 46	„	Pass.	...	...	11 12	...	11 20
p.m.	Mixed	...	...	12 52	...	1 2	p.m.	Pass.	...	...	12 12	...	12 20
„	Engine and van	...	...	...	1†50	1 52	„	Freight	...	...	1 18	1‡20	...
„	Pass.	...	...	2 42	...	2 51	„	Pass.	...	...	2 14	...	2 22
„	Engine	...	NSQ	...	3 23	3 25	„	Engine	...	NSQ	3 0	3‡ 2	...
„	Engine	...	SO	3 42	...	3 51	„	Freight	...	SO	3 0	3‡ 2	...
„	Pass.	...	B	5 0	...	5 9	„	Freight	...	...	...	3 17	3 27
„	Pass.	...	...	5 53	...	6 2	„	Pass.	...	...	4 20	...	4 28
„	Pass.	...	...	6 24	...	6 33	„	Pass.	...	...	5 37	...	5 45
„	Engine	...	NS	7 55	...	8 4	„	Pass.	...	...	6 8	...	6 16
„	Pass.	...	...	9 0	...	9 9	„	Freight	...	NFS	{ 6 43 / ...	6‡45 / 7† 5	7 15
							„	Freight	...	FO	{ 6 52 / ...	6‡54 / 7†10	7 20
							„	Pass.	...	...	8 40	...	8 48

† Depart.   ‡ Arrive.   A—Runs as "Mixed" train when required for the conveyance of live stock only.   B—Horsebox, etc., traffic not to be conveyed by this train.

Above: The summer timetable for the Chard branch taken from the SR Working Timetable for 1932 and worked by the GWR.

Chard Town-Chard Junction

Rail services were often the enterprise of a railway company, but they also came about from pressure generated by a local community. Such was the case with Chard. The town, which had been a borough since 1234, had become by the mid-19th century a thriving market town with a population of approximately 6,000. Local industries included shoe making, Gifford & Fox's line making factory, plus a thriving wool industry which by the mid-1850s gave way to the production of net, thus forming the basis of machine-made lace. The Bridgwater & Taunton Canal opened in 1827 and had become a success, and three years later the businessmen of Chard commissioned a

survey from an engineer, a Mr James Green, the County Surveyor for Devon (who had been involved with the construction of the Bude Canal), regarding the possibility of constructing a canal from Chard to link up with the Bridgwater & Taunton near Creech St. Michael. Having made the survey, the engineer in his report favoured the idea of a railway. However, his advice was disregarded and the decision to build the canal was taken following Parliamentary assent in 1833. Construction was not as fast as had been hoped and several Acts of Parliament were required before the canal, linking the town with the Bridgwater & Taunton from Creech St Michael via Ilminster, was

Below: An early view of the original LSWR branch terminus at Chard Town. Opening on 8 May 1863 and known as the 'tin station' due to its corrugated iron structure, it closed to passenger traffic on 30 December 1916. The GWR thence operating passenger services between Chard Joint and Chard Junction from 1 January 1917; the former LSWR terminus then reverted to goods traffic. *Lens of Sutton*

Above: Adams 'O2' class 0-4-4T No 222 — fitted with a Drummond stove-pipe chimney and strewn with flags — poses for the camera at Chard Town with a special train of road-making vehicles produced by Phoenix Engineering of Chard. The vehicles were destined for road construction works in India in preparation for the Durbar in 1911. *R. Lacey*

opened on 28 May 1842. It traded for 25 years, although traffic never reached the company's expectations and it closed on 29 September 1866, being purchased for £6,000 by the Bristol & Exeter Railway.

The broad gauge Bristol & Exeter Railway had reached the county town of Taunton on 1 July 1842, passing about 10 miles north of Chard, thus encouraging great interest in local railways. With railway mania reaching fever pitch proportions at times, nearly every village and town wanted to be part of the new age, and plans were being mooted to construct railways far and wide. A scheme was proposed in 1847 with the support of the LSWR to connect Yeovil, Dorchester and Exeter and if this had happened, the town of Chard might have had a railway much earlier than 1863. However, the Bill failed due to the opposition of the Great Western Railway which in 1846 was proposing to extend the Berkshire & Hampshire Railway from Westbury via Yeovil, Honiton and Exeter — the title Exeter Great Western was even suggested for this new venture which, indeed, would have shortened the journey from Paddington to Chard. But alas, much to the dismay of the inhabitants, this plan also foundered due to the objections of the LSWR and the B&E.

Other railway projections at the time (1845) included the Bristol and English Channels Direct Junction Railway. This was a scheme to connect Watchet with Bridport via Taunton, Hatch Beauchamp, Ilminster, Chard, Crewkerne and Beaminster. At this time the Bridgwater & Taunton and Chard Canal companies, having seen their trade almost grind to a halt with the success of the B&E at Taunton, now proposed their own rail link known as the Bristol and English Channels Connections Railway and Harbour, this plan involving building a new harbour at Stolford (near Bridgwater on the Somerset coast) and connecting to the B&E, then running alongside the canal as far as Creech St Michael. From here a branch would have continued to Taunton, with the main route following the Chard Canal to Ilminster before continuing to Bridport via Crewkerne and Beaminster, with branch lines to Chard, Axminster and Lyme Regis. Neither of these schemes was successful; however, the Chard Canal Company proposed a line under an Act of 1847 between Taunton and Chard but the waterway company was so heavily in debt it could not afford to construct a railway and, despite a revision by the local populace in 1852 and a further Act in 1855 proposing to convert the canal into a railway, nothing happened.

In 1856, at another meeting in Chard, neither the LSWR nor the B&E was interested in a scheme proposed by the local populace to construct a railway between Chard and Taunton. The local businessmen decided that if a line was to be built, then it would have to be at their own instigation, and at a public meeting held in Chard in November 1859 to discuss the need for a railway, it was concluded that 'in the opinion of this meeting a railway from the London & South Western Railway, with a tramroad to the Canal Basin is of great importance to the prosperity of the town.' Thus started the embryo of a railway between the future LSWR main line station at Chard Road and the town, to meet the need for speedy transportation of passengers and market goods, and also to bring in raw materials for industry. After the election of a committee and receiving a favourable response from the inhabitants, the businessmen instructed their solicitor to prepare an application to Parliament and form a company known as the Chard Railway. From contemporary newspaper reports it appears there was an element of impatience by the inhabitants to get the railway running; the South Western at first objected to the Chard Railway scheme but after a visitation by the local directors to the Board of the South Western, objections were withdrawn after suitable terms for working the line were agreed with the LSWR. Parliamentary Assent was granted on 25 May 1860, and it was estimated the line would cost £25,000 to construct. Meanwhile, the main line of the LSWR had opened from Yeovil Junction to Exeter on 19 July 1860, with its station known as Chard Road opening three miles south of the town, and by August 1860 the scheme for the branch was beginning to take shape. On 14 August, following advertisements being placed by a London solicitor, Mr W. T. Manning, a Mr Hattersley had his tender to make the line accepted. However, the first contractors engaged to construct the line, Messrs Payne & Furness, found themselves in financial difficulties and were released from their contract. The directors, after being unable to agree conditions with a contractor who offered the best price, eventually picked a Mr Taylor. As part of the agreement with the South Western the contract was to be overseen by Mr Galbraith of the aforesaid company, and he was appointed engineer to the Chard Railway Company in place of Mr Burke who was asked to resign. The ceremony of cutting the first sod was undertaken by Miss Susan Buckland on 1 November 1860, with construction starting the following day. However, it was not long before insolvency loomed again when Taylor's prediction of completing the line by 1 June 1861 backfired and, falling foul of his creditors, he had to cease working on the line.

Yet again, the inhabitants of Chard were left out on a limb due to the financial state of the contractors, while the directors of the Chard Railway were entering negotiations to amalgamate with or sell out to the South Western. In early 1861, at the George Hotel in Chard, a committee of local businessmen complained that if the LSWR did not proceed to their satisfaction, then the directors of the Chard Railway would make enquiries elsewhere. This immediately ruffled the feathers of the South Western and stung them into action, although the resolution to purchase the local company for an estimated £20,000 was not passed by the LSWR directors until 30 May. They effectively absorbed the Chard Railway at the end of March 1861 and finally dissolved the Chard company in 1864 under powers obtained by the Act of 22 June 1863. With the South Western now supporting the branch, shares in the Chard Railway were sold by the local directors to some of the top brass of the LSWR including Charles Castleman, the Hon Ralph Dutton, Edward J. Hutchins MP, H. C. Lacy and Wyndham Portal. With the financial clout and construction expertise of the South Western, the way ahead for the line was now set, and by 20 April 1861 Messrs Symonds & Son and Mr Hull, together with Messrs Drew & Son (surveyors for the South Western), began valuing the land, meeting on Monday, 29 April 1861 to settle the purchases. Work was rapidly proceeding, and by 21 February 1863 the line from Chard Road was nearing Chard, while at the lower part of the town the temporary station and goods shed were being erected. It was expected that the line would be ready for traffic by May, and it was reported on 4 April 1863 that the permanent way had been laid and the station house nearly completed; goods shed and coal yards would soon be ready, and there was no doubt that the line would be open for passengers by 1 May 1863. The prediction of 1 May was not far out — they missed it by just seven days, with the line opening on Friday, 8 May 1863, just over three years after the Chard Railway Bill had finally been passed by Parliament.

Even before the opening of the LSWR branch, the local populace was determined to have a railway connection to Taunton, and on 18 March 1861 a meeting was held at the Guildhall in Taunton with Messrs Edwards, Beadon and H. Sutton, Bailiffs to the Borough, presiding. There was a large attendance, and Mr Conybears, the engineer, said there was a very important consideration connected with the line between Taunton and Chard; namely that it would open a communication between the Bristol Channel in Bridgwater Bay and the English Channel at Axmouth Bay. Other people addressing the meeting included a Mr Lee Lee, a prominent local businessman, who retorted that a railway from Chard through the parish of Ilminster to Taunton was an object of great importance to the district and deserved the support of the town of Taunton. At a similar meeting held in Ilminster a few days later similar comments in support of a line from Taunton were freely given, including a statement from Mr Lee Lee reporting he had heard during the day that the LSWR had purchased the Chard Branch which would be connected to them. Thus there would be a through connection from line to line and channel to channel. He also reported that the Bristol & Exeter had agreed to work the line (the

Above: Despite being closed to passenger traffic for 44 years, the old LSWR terminus building at Chard Town looks in a remarkably good condition on 30 April 1960. *A. E. West*

Chard & Taunton) and pay a minimum rate of 3¾% for money invested. The Chard & Taunton Railway was given powers by an Act of 1861 to construct a line between the two towns; again the LSWR and B&E were given the chance to study the proposals, but this time the B&E, ever fearful that its old adversary the South Western would strike out towards Taunton, took up the proposal. Grand ideas are not realised in the hard commercial world without the backing of financial investment and, despite the issue of new shares to the value of £2,000, there were not enough local people willing to invest in the Chard & Taunton. The London money market was currently charging interest rates of 8 or 9% at the time and, as a result, the B&E was unable to begin work immediately. Thus an extension act was required in 1862, culminating in the B&E gaining control of the Chard & Taunton with an Act of 1863, and enabling it to build the line to its own broad gauge standards. Construction started in 1864, with trains entering the branch from Creech Junction on the GWR main line.

There are reports of a strike for more pay by the navvies engaged on construction of the new line in April 1865, but by 20 May of the same year extra navvies were being employed to work on the line. The bridge near the LSWR station was nearly completed, and great progress throughout the whole line was in evidence (the bridge referred to was the road bridge carrying the present-day A30 road over the LSWR spur line to the new joint station). By 7 July 1866 the permanent way was nearing completion, the townspeople of Chard were looking forward to an opening on 1 August 1866 but the directors of the line passed through on 4 August and expressed satisfaction with the work done, although the Board of Trade Inspector had yet to examine the line. This was completed successfully on 4 September and the line opened to passenger traffic on 11 September. (Freight services did not start until March of the following year.) However, the Inspector did not grant the same satisfaction to the LSWR for its connecting spur which required considerable alterations, as the LSWR was optimistically trying to make use of its existing tram line running from Town station to the canal basin which was the site of the new joint station. This tramway had originally been authorised in the Chard Railway Company Act to run from the South Western station to the canal basin and to be worked by horse power only. The link was upgraded for passenger traffic and came into use on 26 November 1866, the spur line connection running from a point just short of its terminus at Chard Town and diving through a short cutting with a steep gradient down to the joint station.

Chard Town

At last the inhabitants of the town were now connected to the railway, but not as they had hoped. Although the branch would be beneficial, it was by no means the answer that the local businesses had hoped for. To transport goods to the South West, Bristol or the Midlands meant sending them to Exeter or Yeovil where the transfer to the broad gauge was time-consuming and costly, as well as involving a very circuitous route. Thus the South Western was able to enjoy complete control of goods and passenger traffic from Chard for nearly a further 3½ years. Celebrations for the opening day started early, with the population being awakened by the pealing of church bells at 5am, and with the departure of the first train to Chard Road timed at 7.40am there was a large and excited gathering on the platform. The inaugural train, containing 20 passengers, was hauled by *Firefly*, a 2-2-2 well tank, which was decorated with flowers for the occasion. With much shouting and cheering from the populace, the small engine started on its 3 mile 110 yard journey to Chard Road at which a new platform had been especially constructed on the opposite side of the station approach from the main building. After awaiting the passage of a main line train it returned to Town station. The Mayor and civic dignitaries were conveyed on the 11.30am to the main line station and returned for the celebration lunch, while the navvies who had built the line held their celebrations in the Red Lion. The initial services in 1863 consisted of six passenger trains in each direction from Monday to Saturdays, plus three trains in each direction on Sundays. The stock consisted of four-wheeled coaches with a formation of a brake van and two or three coaches. The station at Chard Town consisted only of a short, single platform, complete with a small station building with a roof constructed of corrugated iron and it was not long before the nickname 'tin station' was applied by the locals. Cattle pens were supplied at the end of the platform nearest the town; a substantial brick-built goods shed and attendant sidings were also provided to serve the local traders, including Messrs Bradford & Sons and the Somerset Trading Company. A signalbox controlled the layout.

Harmonisation of the two companies was further realised in 1896, when the LSWR terminus at Chard Town and the adjacent spur platform came under the control of the GWR stationmaster of Chard Joint. The South Western kept its own booking clerk at Town station, while the GWR took charge of the permanent way between the junction and joint stations. Passenger services, worked by the South Western, were not a viable proposition; although the company was working services to the joint station, it still retained a service to the original terminus at the town station and by 1916 it was realised by both companies that for economic reasons, especially with the conditions brought about by wartime austerities, they would have to meet and discuss the situation. The outcome was that the GWR would now work the whole branch to Chard Junction, with the South Western terminus at Chard Town and the spur platform closing to passengers from 30 December 1916. The signalbox was closed from the same date and a ground frame installed, opened by the key on the end of the single-line tablet, enabling goods trains to gain access to the old station which became the main goods depot for Chard and district. The South Western coaching stock and locomotive, which had been based at the former terminus for working the branch trains, were withdrawn leaving the GWR in complete charge of services, although LSWR locomotives continued to convey goods wagons from Chard Junction to the former terminus for a few years.

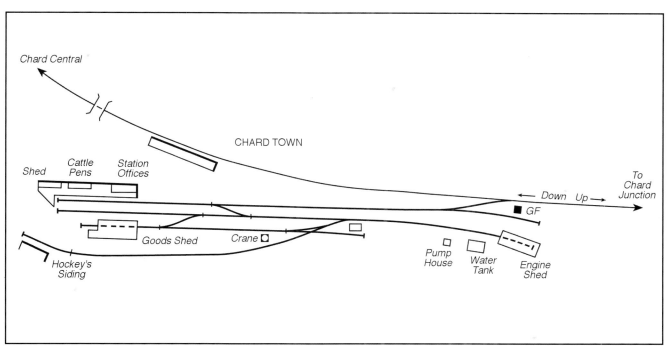

Above: A view of the original Chard Town terminus closed on 30 December 1916 in use as a goods depot on 12 April 1953. Heaps of coal adorn the platform where passengers once awaited the train to Chard Road. *Adrian Vaughan collection*

Chard Town Locomotive Shed

A small engine shed, measuring 50ft by 20ft, complete with a curved roof, was supplied to house the branch tank engine. A coal stage and enginemen's cabin were situated outside the shed, but no turntable was installed as only small tank engines were used on the branch. As mentioned, the locomotive *Firefly* worked the inaugural service between Chard Road and Chard Town; this locomotive, a 2-2-2 well tank was assembled at Nine Elms between November 1862 and 27 February 1863 when it emerged from the works using the boiler, frames and leading wheels from an earlier locomotive, namely a Rothwell-built 2-2-2 tender engine No 76 *Firefly* dating from February 1847 which had been withdrawn in November 1862. The 'new' *Firefly*, when completed, had 3ft 6in leading wheels, 5ft 6in driving wheels and 4ft trailing wheels; the engine had a weight of 20½ tons and a water capacity of 210 gallons. Another 2-2-2WT, No 13 *Orion* of the 'Tartar' class, built by

Sharp Roberts in 1852, was at work on the Chard branch in 1863, becoming the regular branch engine until being withdrawn in April 1872. This class of well tank had the distinction of having, at 6ft, the largest driving wheels of any LSWR tank locomotive. At a later date, No 9 *Chaplin*, another 2-2-2 well tank, was employed on the branch. The allocation for March 1878 shows No 207, a standard 2-4-0 well tank at Chard, with No 218 of the same class being there in March 1890. The Beattie well tanks were used up until the turn of the century, with an engine being outstationed from Yeovil for the branch trains. An Adams 'O2' 0-4-4 tank is reported as being allocated to Chard Town in March 1900.

The LSWR worked trip freights between the junction and town stations for a few years after the GWR took control of the passenger services and, on occasions, one of the 'Ilfracombe Goods' 0-6-0s worked the branch, with appearances by the Adams Radial tanks. The summer timetable of 1909 shows a weekly excursion to Seaton from Taunton comprising GWR stock, the train being worked on from Chard Junction by a South Western locomotive sent down especially from Yeovil Town. However, the LSWR

Above: The branch terminus at Chard Junction comprised a single platform complete with attractive canopy. This was the starting point of the first branch line to Chard; the terminus at Chard Town was opened by the LSWR on 8 May 1863. The main line station building can be seen to the right. *R. Lacey*

1914 Excursion Train Special Notice shows the GWR engine working through to Seaton with an LSWR pilot driver being provided by Yeovil shed. The formation of the Southern Railway at the Grouping in 1923 witnessed the Chard shunting locomotive returning to Yeovil each evening, the locomotive shed at the former terminus being removed in 1929 although it had been in commercial use since 1920.

Chard Junction

Opened with the main line as Chard Road, the station was renamed Chard Junction in 1872. The South Western branch passenger services totalled 11 return workings by 1897, with four of these services running to and from the joint station to connect with the GWR services. The rest terminated at Chard Town, with one mixed train between the junction and joint stations, plus an early morning and evening freight train. Sunday workings had now ceased. The GWR started operating passenger services to Chard Junction from 1 January 1917 with a separate set of coaches but, after a time, it was deemed more economical if trains

between the two stations were worked as an extension of the Taunton to Chard services. With the suggestion in late 1907, that a railmotor service should be provided between Chard Junction and Crewkerne, the inhabitants of Winsham near Chard convened a meeting in their Jubilee Meeting Room in February 1908 to make clear to all that a halt must be provided in the neighbouring village of Clapton but as there was no road access between Winsham and Clapton this would seem to be an ambitious request. Also at the same meeting, another ambitious scheme received its first airing when it was announced that Hawkchurch Parish Council wanted nothing less than an extension of the railmotor service from Honiton to Yeovil Junction and their district served by halts at Winsham and Hewish. The scheme was placed before Mr G. F. Vallance — the LSWR District Superintendent at Exeter — and he met the local representatives at Chard Junction in March 1908. His promise to refer the matter to the South Western Board must have been a public relations gesture, the directors announcing their refusal of the plan in May of the same year. The reason given was insufficient traffic, as train operation in this sparsely populated

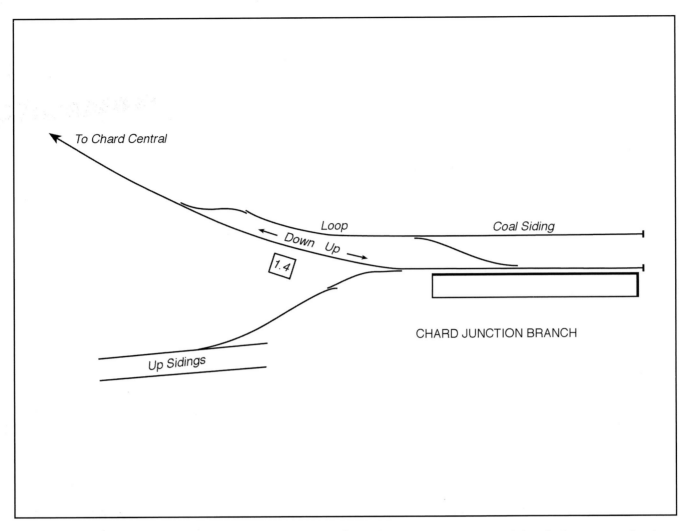

To Chard Central

Loop   Coal Siding

Down   Up →

1.4

CHARD JUNCTION BRANCH

Up Sidings

area would not have been in the best interests of the South Western. Additionally, the railmotors themselves would not have been suited to a journey length of 40 miles, with their slender resources of coal and water and, more importantly, no railmotor would have had the power to travel up Honiton bank from Seaton Junction.

An interesting working occurred during the two summers before World War 2. As there was no GWR Sunday service to Taunton, apart from a midday milk train, the Southern Railway operated a Sunday service all day between Chard Junction and Chard for excursion passengers. This service was worked by a Southern Railway engine and a two car push-and-pull set sent down from Yeovil, with the engine and stock returning in the evening. This was the only known time that the LSWR or the Southern Railway had worked a passenger service between Chard Junction and Chard since the closure of Chard Town in December 1916. Chard Junction was unique in the fact that its branch trains did not operate from a bay platform, as at other locations on the route, but from a separate platform situated opposite the main station building. The single platform was adorned with an attractive canopy that was removed in its final years. A small run-round loop, leading through a siding used by local coal traders, was used by the engine to run around its coaches before returning to Chard Central

and Taunton. At the end of the platform a set of points led to the left making a trailing connection into the main line up goods yard; this was the actual junction between branch and main line and an awkward one to operate, especially when the excursion trains arrived from Taunton. It was the custom on the two summer Bank Holiday Mondays at Whitsun and in August to run an excursion train from Taunton, calling at all stations on the branch, to either Lyme Regis or Seaton. The trains were well patronised, especially if the weather was good; however, due to the restricted layout, trains could not be shunted in one whole movement, and had to be split for the onward journey via the Southern main line, with the whole process having to be repeated on the return to Taunton. The excursion engine did not work beyond Chard Junction as it had to return to Chard Central with the train staff to enable other trains which formed the standard service to operate. Excursion trains ran in BR days on August Bank Holiday Mondays, with the 10.15am ex-Ilminster to Seaton (six coaches) and the 10.37am Taunton-Seaton (six coaches). A Yeovil 2-6-0 worked the Ilminster train as far as Seaton Junction, returning light engine for the 10.37am ex-Taunton while the two evening return trips were similarly worked between Seaton Junction and Chard Junction by the same locomotive. The final excursions ran in August 1962.

Above: An Edwardian summer view of Chard Junction with a gathering of passengers on the down platform to the left. Advertisements adorn the building on the up platform including Maple & Co, Players Navy Cut cigarettes and Pears Soap. *R. Lacey*

Viewed from the single platform, the branch rose at 1 in 80 and curved into the distance. A small ground level signalbox was situated just past the connection into the goods yard. At one time there was signalling on this portion of the branch, but this was removed for economic reasons on 5 March 1935 and the branch box reduced to ground frame status, the section between Chard Junction and Chard being converted to 'one engine in steam' working. When trains left Chard Central, the signalman there telephoned his opposite number at Chard Junction to inform him of departure. A train staff was provided as the driver's authority to occupy the line, and this embodied a key to the Chard Town goods yard. Similarly, the signalmen at either end were informed by telephone when work at Chard Town was complete. The branch box was taken out of use on 28 July 1964 and the points converted to hand operation.

As one would expect, the branch platform with its two-coach 'B' set passenger train headed by a Taunton pannier tank or Prairie tank locomotive gently steaming away, was a far cry from the main line station with trains thundering through on their way to and from Waterloo. Some fine running could be witnessed, especially with trains heading down from the summit at Crewkerne Tunnel and, with whistles screaming they would slam over the level crossing heading towards Exeter and Plymouth. The nearby rail-connected United Dairies complex-handled vast amounts of rail-borne milk traffic; much shunting was undertaken here making this one of the major distribution centres on the Southern for the nightly move of milk to London. Chard Junction closed to passenger services on 7 March 1966 and to freight on 18 April of the same year. The milk factory sidings are still retained but remain unused. The station buildings have long gone but the goods shed still survives and the branch platform now forms part of a coal yard. The LSWR signalbox, dating from 1875, survived until September 1982 when it was replaced by a modern building which now controls the important crossing loop on the single-line section between Yeovil Junction and Honiton. Of Chard Town there is no trace as a supermarket now occupies the site, but the former joint station building at Chard Central still survives.

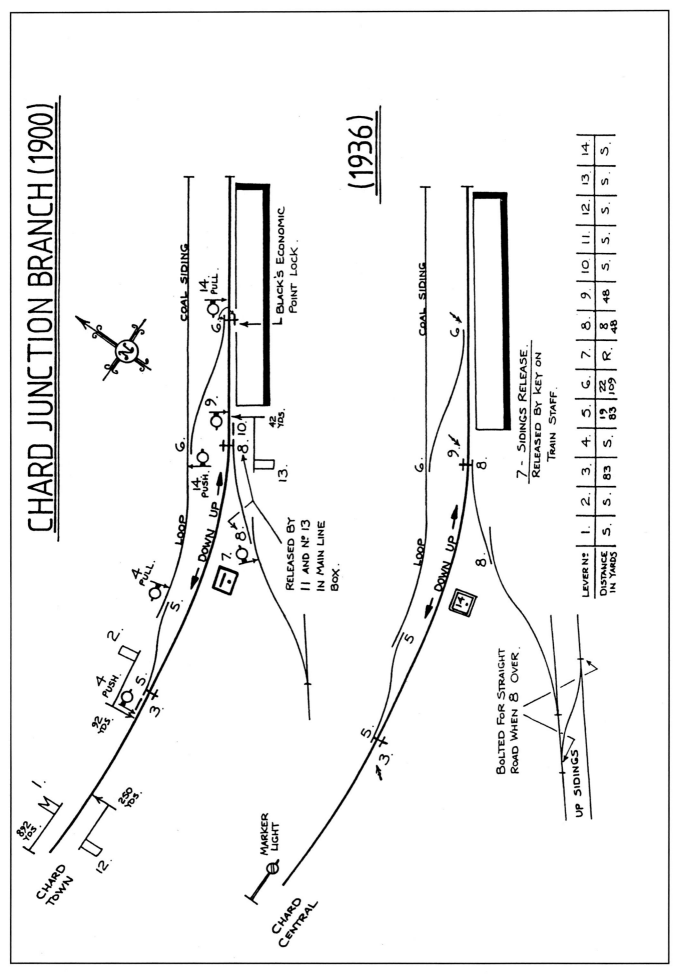

CHARD JUNCTION BRANCH (1900)

(1936)

CHARD TOWN

CHARD CENTRAL

COAL SIDING

BLACK'S ECONOMIC POINT LOCK.

RELEASED BY 11 AND Nº 13 IN MAIN LINE BOX.

MARKER LIGHT

BOLTED FOR STRAIGHT ROAD WHEN 8 OVER.

UP SIDINGS

7 - SIDINGS RELEASE. RELEASED BY KEY ON TRAIN STAFF.

LEVER Nº	1.	2.	3.	4.	5.	6.	7.	8.	9.	10.	11.	12.	13.	14.
DISTANCE IN YARDS	S.	S.	S.	83	19 83	22 109	R.	8 48	8 48	S.	S.	S.	S.	S.

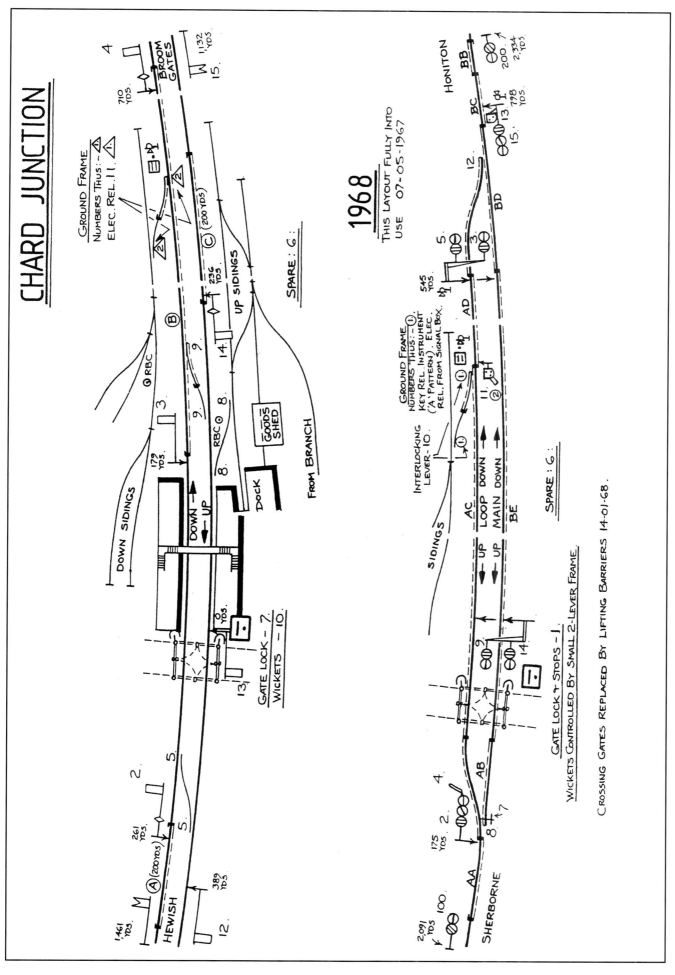

CHARD JUNCTION

GROUND FRAME
NUMBERS THUS :- 11.
ELEC. REL. 11.

SPARE : 6 :

1968

THIS LAYOUT FULLY INTO
USE 07-05-1967

GATE LOCK - 7.
WICKETS - 10.

FROM BRANCH

UP SIDINGS

DOWN SIDINGS

GOODS SHED

DOCK

DOWN ↓ ↑ UP

HEWISH

BROOM GATES

GROUND FRAME
NUMBERS THUS :- 1.
KEY REL. INSTRUMENT
('A' PATTERN). ELEC.
REL. FROM SIGNAL BOX.

INTERLOCKING
LEVER - 10.

SPARE : 6 :

GATE LOCK & STOPS - 1.
WICKETS CONTROLLED BY SMALL 2-LEVER FRAME.

SIDINGS

AC ← UP LOOP DOWN
BE ← UP MAIN DOWN

SHERBORNE

HONITON

CROSSING GATES REPLACED BY LIFTING BARRIERS 14-01-68.

55

Above: Drummond 'F13' class 4-6-0 No 331 rushes through Chard Junction towards Exeter. As well as carrying the normal discs for the route, the locomotive is also displaying the disc code for a special train — a white disc with a black centre over the nearside buffer. The disc would be replaced at night by a purple light. No 331, dating from September 1905, was condemned on 15 February 1924. *Lens of Sutton*

Below: Drummond 'Greyhound' 'T9' 4-4-0 No 304, in superb external condition and only two years old, runs over the level crossing at Chard Junction in 1902 with a Waterloo to Exeter express. Dating from December 1900, the locomotive had been allocated to Exmouth Junction in mid-1902. *R. Lacey*

The approach to the branch platform at Chard Junction with the run-round loop to the left. The connection to the goods yard and consequently the main line can be seen passing to the right. The Chard Junction branch signalbox was reduced to a ground frame on 5 March 1935 when the signalling was removed between Chard Junction and Chard Central. The box was eventually abolished on 28 July 1964. *R. Lacey*

Above: Class 5700 0-6-0PT No 9671 leans into the curve while approaching Chard Junction from Chard Central on 3 August 1957. A temporary speed restriction of 15mph applied to Taunton-bound trains. *A. E. West*

Above: The branch platform at Chard Junction on 7 July 1962 with No 3736 waiting to depart with the 7.26pm to Taunton. The main line station building can be seen in the background. *H. C. Casserley*

Below: Bulleid 'Merchant Navy' class 4-6-2 No 35016 *Elders Fyffes* stands on the level crossing with an up service at Chard Junction on 3 August 1957. *A. E. West*

59

Above: Chard Junction, looking towards Crewkerne on 2 March 1956. The main station buildings situated on the up platform can be seen to the left, just past the footbridge. *R. M. Casserley*

Chard Joint Station

Provided with a fine all-over Brunelian roof, the station and its layout was a more grandiose affair than the facilities at Chard Town. The B&E used the north bay, while the South Western used the south bay with a fine canopy for the benefit of its passengers. The bay was initially no more than the length of three of the carriages of the period. The B&E laid sidings and constructed their goods shed near the now defunct canal basin, and a mixed gauge line ran through the station alongside the interchange platform which was used for the transfer of goods between the two gauges. The only other mixing of the differing gauges was in connection with the turntable which was jointly owned by the two companies and, as was often the case with joint railway ownership, each company had their own stationmaster and staff, their own signalboxes and booking offices. The broad gauge company provided five trains each way on weekdays, and two each way

on Sundays, and also had a locomotive depot for their locomotives. Some of the South Western services from Chard Road ran direct to the joint station to make connections with the B&E Taunton trains, thus effectively bypassing Chard Town, while trains still using the South Western terminus had to reverse out before resuming their journey to the new station. The LSWR constructed a platform complete with a shelter on the connecting spur line lying adjacent to Town station, for passengers using their services and running direct to the joint station from Chard Road. This new platform was unstaffed and tickets were issued from Town station. There were indications that the South Western had ideas of closing the old terminus and constructing a new station on the spur platform, but this did not come to fruition. The GWR gained control of the Bristol & Exeter on 1 January 1876 with the amalgamation of the two companies taking place from 1 August the same year. In 1878 the GWR's General Manager, James Grierson, strongly recommended to the Board that all broad gauge branches, save that to Chard,

Above: The joint LSWR/GWR station at Chard viewed from the station approach *c*1912. The canopy to the right covers the LSWR platform for the benefit of passengers awaiting the train to Chard Junction. *LRGP*

Below: An overall view of the impressive station building at Chard Central on 14 April 1953. The square-post ex-GWR lower quadrant signal in the foreground stands at danger, guarding the route to Chard Junction.
Adrian Vaughan collection

Above: The 1.35pm to Taunton formed of two-coach 'B' set Nos W6999W and W7000W and headed by '5700' class 0-6-0PT No 4604 waits expectantly for passengers under the overall roof of the former Bristol & Exeter and LSWR joint station at Chard Central on 2 July 1956. *Hugh Ballantyne*

should be converted as soon as possible. The fear was still there that the LSWR would obtain compulsory running powers to Taunton if the former B&E branch was made 'narrow gauge'. The GWR Wilts, Somerset & Weymouth line had been converted on 18-25 June 1874, and by June 1884 the broad gauge had disappeared from 108 more miles of railway, 25 of which had been previously mixed. The Chard branch was now the only broad gauge, as opposed to mixed gauge line, east of Exeter. Seven years later the branch succumbed to the inevitable.

The conversion of the gauge on the GWR branch made things less difficult so far as interchange of goods, etc, between the two companies was concerned, although traffic on both branch lines was never heavy. The Bristol & Exeter used its Pearson 4-4-0 saddle tanks in the early days, from Taunton to the joint station, and after the B&E was absorbed by the GWR in 1876, a 'Hawthorn' class 2-4-0 tender engine was used. With two kept at Taunton for working the branch, one would have been based at Chard shed for a week at a time. The GWR allocation for Chard on 1 January 1901 was an 0-4-2T, No 544 and an 0-6-0 saddle tank, No 1145. The allocations for

1903 were 2-4-0 'Metro' tank No 3590 and 0-6-0ST No 1641. The 'Metro' tank is recorded as working the whole of the passenger service between Chard and Taunton in 1903 with the exception of the 9.55pm passenger and mail, and the return mail service from Taunton at 3.15am, as these services were worked by No 1641. The last two engines allocated to the GWR shed at Chard in 1924 were 0-6-0PT No 1575 and 2-4-0 'Metro' tank No 1446. The loco shed closed on 14 July 1924, locomotives then operating from the parent shed at Taunton. From the 1930s the GWR services were in the hands of the '55xx' 2-6-2 Prairie tanks hauling B set bogie coaches, and later '5700' pannier tanks. BR Standard 2-6-2 tanks appeared alongside the ex-GWR locomotives after Nationalisation. The tracks leading into the south bay at the joint station, previously used by South Western trains from Chard Road, were removed on 10 October 1927; Chard South signalbox was closed in 1928, and the word 'Joint' was dropped from the station name from 1 March of the same year with the station becoming 'Chard'. The turntable was removed in November 1935, there being no longer a use for it as the branch was worked mainly by tank engines.

Above: Chard Central as viewed from the Taunton end on 7 July 1959. The station building still exists today in use as commercial premises. *R. M. Casserley*

Upon Nationalisation in 1948, the two original sections of the branch were allotted to the Southern and Western Regions, with the WR operating the whole of the branch. Chard became Chard Central from 26 September 1949, and the Southern Region gained control of the line from Chard Junction to Thornfalcon in 1950, although the operation of the line was left to the Western Region. Then came a fuel crisis which gripped the whole country; the branch was closed from 3 February 1951 and for around five months Chard was without a railway. It was the beginning of the end; the inhabitants of Chard found it was actually more convenient to board a bus in the High Street and travel to Taunton rather than to walk down to the station at the bottom end of the town. Many people believed that the train services would never be reinstated, however, after strong protests the line was reopened to passenger traffic on 7 May 1951 and continued to serve Chard for another 11 years.

Passenger traffic on the branch was light, with 25,901 tickets issued at Chard in 1903, reducing to a total of 7,434 in 1933. A census taken in July 1961 revealed a daily average of fare paying passengers

alighting from all stations, including Taunton and Chard Junction, was only 155. An average of only four passengers alighted from each of the daily services arriving at Chard Central; the total average of passengers leaving Chard Junction for Taunton by the seven daily trains was only 32. It had been suggested that the introduction of diesel units, plus the abolition of signalling and leaving the stations unstaffed, would result in savings on the branch, but a statement from BR explained that the 'introduction of new diesel units would not effect an improvement to an extent whereby the service could be made remunerative, and it would be unreasonable to expect the Commission to incur a capital expenditure in the provision of these units when there would be no hope of any return.' The branch was now doomed. The case was presented in 1961 by British Railways, subject to the agreement of the South Western Transport Users Consultative Committee, to withdraw passenger services between Taunton and Chard, and Chard and Chard Junction. The loss of revenue from the proposed closure was expected to be less than £4,000 per year, while working expenses alone amounted to £22,500. The closure would result in an annual net saving of

£18,000, this not taking into account the cost of renewing permanent way, signalling and bridges. The passenger service was withdrawn from Monday, 10 September 1962, the final trains running on the previous Saturday with pannier tank No 4663 heading the last train from Taunton, the 6.50pm, in the hands of Driver Millard and Fireman M. Chandler of Taunton shed who were given an official send-off by Mr E. Frampton, the Taunton stationmaster. Watched by a considerable crowd, the pannier tank, with over 120 passengers on board the train, steamed out from Taunton amidst exploding detonators, loud cheers from the onlookers, and long whistle blasts from other locomotives standing in the vicinity. Upon arrival at Chard Central, the driver was presented with a bunch of dahlias, and many more people boarded the train for the short run to Chard Junction where other passengers awaited to travel on the final train of all, the 9.30pm to Taunton, which left 15 minutes late due to a late connection with a main line train. No 4663 carried express headlamps and, in time-honoured fashion, pulled away with much whistling and exploding of detonators.

This was not to be the last ever passenger train to traverse the branch however, this accolade belonging to the 'Quantock Flyer', a rail tour organised by the LCGB which ran over the line on 12 February 1964 headed by 2-6-2T No 4593 and 0-6-0PT No 9663. The route taken by the rail tour was Waterloo-Yeovil-Taunton-Chard and return to Waterloo. The train was hauled by a 'Merchant Navy' 4-6-2 from Waterloo to Yeovil Junction, and the two WR tank engines took the 10-coach special via Yeovil Town and the old B&E branch to Curry Rivel Junction and thence to Taunton, whereupon they ran round and hauled the train to Chard Junction via Creech Junction. At Chard Junction the train had to be divided five times in order to be shunted over the connecting line into the up siding where No 35030 *Elder-Dempster Lines* was waiting to haul the railtour back to Waterloo.

General goods traffic was withdrawn from Chard Central on 3 February 1964 and the line north of the station was closed on 6 July 1964, being lifted from Creech Junction to Chard Central by 6 August the same year. Thus ended the railway from Taunton. The railway had not left Chard completely however, as the original London & South Western line from Chard Junction to Chard Central was used as a long siding and remained open for goods — tankers of molasses for B. G. Wyatt Ltd, and tankers of tar oil for Lion Emulsions at Chard Central. After an absence of many years, footplatemen from Yeovil Town shed regained the right to work trains along the former LSWR branch between Chard Junction and the former joint station after having been worked for so many years by the GWR and the WR. Unfortunately, by 1966 the remaining trade had almost ceased. With the goods yard at Chard Town closing on 18 April 1966, the remainder of the branch was closed on 3 October and was lifted by December 1967.

Footplate Memories

The branch was well known for two remarkable runaway accidents that happened at Chard Junction. The first was during the 1930s when a 2-6-2 tank and two coaches heading down the bank from Chard Central ran through the stop blocks at Chard Junction and out into the main road. The second, and perhaps the most dramatic incident, happened during the 1950s when Taunton footplatemen Driver Reg (Nimble) Manning and Fireman Merv Chandler were shunting at Chard Town. I will now let Merv describe what happened.

'We were on the 9.10am Chard Goods one day, with Don Pavey as guard. While shunting at Chard Town we had to shunt the brake van and three wagons back towards Chard Junction when we were forming the train to return to Taunton. As the wagons hit the brake van it started to move. We shouted to Don who ran after them, but to no avail as they kept just ahead of him. We could see him running for a long way, but after a while he came back and said that he couldn't catch up with them, and the last he saw of the wagons they were starting down the gradient towards Chard Junction. It was all down hill, and how they stayed on the curve with the check rail I will never know. We found out later that they had gone into Chard Junction, smashed up the blocks, crossed the main road and had gone up the driveway of the pub.' Witnesses recall that the 20-ton brake van and the six-plank open wagon containing machinery were travelling at an estimated 45mph as they sped down the 1 in 80 gradient and rumbled alongside the branch platform, snapping the fishplates and pushing the stop blocks through the fence. The wagons then careered across the road and came to rest in the car park next to the lounge bar of the Chard Road Hotel. It was very fortunate that the level crossing gates were open to road traffic at the time as the outcome would otherwise have been far more serious, but by great fortune no casualties occurred. However, the landlord of the inn did not seem too concerned, as that evening he had one of the best nights' takings for some time, and in fact the pub was virtually drunk dry. A photograph of the incident appeared on the front page of the *Chard & Ilminster News* the same week, again adding to the publican's takings.

My own firing days at Chard Junction involved working on freight trains and the local stopping trains out of Yeovil Junction and Exeter Central and I have described these workings in my previous books. One incident severely dented my ego when working as a young fireman on a stopping passenger train from Exeter Central to Salisbury, with a relief crew awaiting us at Yeovil Junction to work the train onwards. This particular train involved a bit of a wait at Chard Junction while mail and parcels etc were being loaded. While en route from Axminster, it was

Above: '5700' class 0-6-0PT No 9670 is pictured at Chard Central after arrival with the 8.30am from Taunton on 2 March 1956. The south bay track used by LSWR trains originally ran to the left and alongside the platform in the foreground before removal together with the awning in 1927. *H. C. Casserley*

agreed that as soon as we ran into Chard Junction I would hop off with my billy can and make a brew from the station kettle and my mate would make up a good fire in preparation for the climb to Crewkerne. We ran into the station at a fair lick; I opened the small door between the cab and tender of our 'S15' 4-6-0 and stood on the top step ready to hop down on to the platform. However, pride comes before a fall, and with a bit of showing off to the passengers awaiting our train, I jumped down. Unfortunately, I landed with my back to the engine and the forward momentum turned me head over heels on the platform, the tea can spinning out of control on the platform with tea and sugar going everywhere. The signalman's face was a picture, with a great big grin, and to make matters worse, we had a trainload of passengers to witness my acrobatic skills. I slunk red-faced into the station to make the tea, with a few bruises to show for my misfortune, but none to match the large bruise to my pride!

LYME REGIS BRANCH.

WEEKDAYS.
Summer Service, 18th July to 10th September only.

TO LYME REGIS.

				Axminster	Combpyne	Lyme Regis
	Distances.			—	4m. 21c.	6m. 59c.
				dep.	dep.	arr.
a.m.	Freight	...	---	7 23	{ 7 35 / 7 45	7 53 }
,,	Pass.	...	...	8 35	8 48	8 55
,,	Pass.	...	...	9 39	9 52	9 59
,,	Pass.	...	NS	10 43	10 56	11 3
,,	Pass.	...	SO	11 0	11 13	11 20
p.m.	Pass.	A	NS	12 25	12 38	12 45
,,	Pass.	...	SO	12 51	1 4	1 11
,,	Pass.	D	NFS	2 35	2 48	2 55
,,	Pass.	D	FSO	3 0	3 13	3 20
,,	Pass.	...	NS	4 12	4 25	4 32
,,	Pass.	...	SO	4 20	4 33	4 40
,,	Pass.	...	...	5 15	5 28	5 35
,,	Pass.	...	NFS	6 35	6 48	6 55
,,	Pass.	D	FO	6 47	7 0	7 7
,,	Pass.	D	SO	6 55	7 8	7 15
,,	Pass.	AB	...	8 50	9 3	9 10

FROM LYME REGIS.

				Lyme Regis	Combpyne	Axminster
	Distances.			—	2m. 38c.	6m. 59c.
				dep.	dep.	arr.
a.m.	Freight	...	...	6 15	{ 6 24 / 6 34	6 45 }
,,	Pass.	...	...	8 5	8 14	8 25
,,	Pass.	H	...	9 10	9 19	9 30
,,	Pass.	...	...	10 5	10 14	10 25
,,	Pass.	E	SO	10 26	10 35	10 46
,,	Pass.	E	NS	11 43	11 52	12 3
,,	Pass.	E	SO	11 58	12 7	12 18
p.m.	Pass.	E	SO	1 48	1 57	2 8
,,	Pass.	...	NS	1 55	2 4	2 15
,,	Pass.	F	...	3 40	3 49	4 0
,,	Pass.	...	NS	4 42	4 51	5 2
,,	Pass.	...	SO	4 48	4 57	5 8
,,	Pass.	...	NS	6 5	6 14	6 25
,,	Pass.	...	SO	6 15	6 24	6 35
,,	Pass.	...	...	8 20	8 29	8 40

A—Runs as Mixed train when required.   B—Must not run as a "Mixed train" on Saturdays.   D—Conveys through carriage from Waterloo.   E—Conveys through carriage for Waterloo.   F—On Fridays and Saturdays, conveys through carriage for Waterloo.   H—Conveys through carriage for Waterloo, Mondays only.

SUNDAYS.
17th July to 25th September only.

TO LYME REGIS.

		Timing No.	Axminster	Combpyne	Lyme Regis
			dep.	dep.	arr.
a.m.	Pass.	...	10 22	10 35	10 42
,,	Pass.	...	11 17	11 30	11 37
p.m	Pass.	...	12 16	12 29	12 36
,,	Pass.	...	1 18	1 31	1 38
,,	Pass.	...	2 40	2 53	3 0
,,	Pass.	...	4 2	4 15	4 22
,,	Pass.	...	5 30	5 43	5 50
,,	Pass.	...	6 30	6 43	6 50
,,	Pass.	...	7 38	7 51	7 58
,,	Pass.	...	8 50	9 3	9 10
,,	Pass.	...	10 30	10 43	10 50
,,	Pass { 24th July, 21st Aug. 18th September. }	303	11 45	11 58	12 5
,,	Pass { 7th August, 4th September. }	303	12 0	12 13	12 20

FROM LYME REGIS.

		Timing No.	Lyme Regis	Combpyne	Axminster
			dep.	dep.	arr.
a.m.	Pass.	...	9 55	10 4	10 15
,,	Pass.	...	10 50	10 59	11 10
,,	Pass.	...	11 49	11 58	12 9
p.m.	Pass.	...	12 45	12 54	1 5
,,	Pass.	...	1 48	1 57	2 8
,,	Pass.	...	3 30	3 39	3 50
,,	Pass.	...	5 0	5 9	5 20
,,	Pass.	...	5 56	6 5	6 16
,,	Pass.	...	7 10	7 19	7 30
,,	Pass.	...	8 15	8 24	8 35
,,	Pass.	...	9 45	9 54	10 5
,,	Empty { 24th July, 21st Aug. 18th September. }	303	11 10	...	11 30
,,	Empty { 7th August. 4th September. }	303	11 30	...	11 50

LYME REGIS BRANCH.

WEEKDAYS ONLY.
Winter Service, Commencing 12th September.

TO LYME REGIS.

				Axminster	Combpyne	Lyme Regis
	Distances.			—	4m. 21c.	6m. 59c.
				dep.	dep.	arr.
a.m.	Freight	...	...	7 23	{ 7 35 / 7 45	7 53 }
,,	Pass.	...	...	8 35	8 48	8 55
,,	Pass.	...	...	9 40	9 53	10 0
,,	Pass.	...	...	10 43	10 56	11 3
p.m.	Pass.	...	A	12 25	12 38	12 45
,,	Pass.	...	...	2 40	2 53	3 0
,,	Pass.	...	...	4 12	4 25	4 32
,,	Pass.	...	...	5 12	5 25	5 32
,,	Pass.	...	...	6 35	6 48	6 55
,,	Pass.	...	A	8 50	9 3	9 10

FROM LYME REGIS.

				Lyme Regis	Combpyne	Axminster
	Distances.			—	2m. 38c.	6m. 59c.
				dep.	dep.	arr.
a.m.	Freight	...	...	6 15	{ 6 24 / 6 34	6 45 }
,,	Pass	...	...	8 5	8 14	8 25
,,	Pass	...	...	9 10	9 19	9 30
,,	Pass	...	...	10 5	10 14	10 25
,,	Pass	...	...	11 50	11 59	12 10
p.m.	Pass	...	...	1 55	2 4	2 15
,,	Pass	...	...	3 40	3 49	4 0
,,	Pass	...	...	4 42	4 51	5 2
,,	Pass	...	...	6 5	6 14	6 25
,,	Pass	...	...	8 20	8 29	8 40

A—Runs as mixed train when required.

Above: Through coaches to and from Waterloo, mixed and freight trains plus the normal branch workings for the Lyme Regis branch as seen in the SR Working Timetable for 1932.

Axminster-Lyme Regis

The ancient borough and medieval port of Lyme Regis on the Dorset coast gained its royal title when King Edward I used its harbour while engaged in his wars with France, and from that time Lyme became a prosperous seaport, at one time fitting out men-of-war to do battle with our enemies including the Spanish. The town was also a centre for the manufacture of serge, as well as having a flourishing sea trade with the Continent. The English Civil War found the town besieged by the Royalists in 1642, and in 1685 the ill-fated Duke of Monmouth landed here, raising his standard in the market place. However, by the 17th century the harbour was in decline as more modern vessels found their increasing size made it difficult or virtually impossible to enter, this lack of trade rebounding on the fortunes of the town. The gradual transformation from trading port to resort started in the 18th century when a local innkeeper purchased one of the then new-fangled bathing machines, and by the early 19th century it was developing as a resort for the well-heeled gentry who could afford to enjoy the picturesque coastline dominated by Golden Cap. At 617 feet high this is the highest cliff on England's southern coast. Rail communication was vital for this attractive harbour town with its massive semicircular breakwater known as the Cobb that extends from the Customs House seawards for a distance of 870 feet, its links with Jane Austen and popularity with fossil hunters, and a rail connection was soon being demanded, but like all early railway schemes, construction would be far from easy.

Among the early schemes were three dating from 1845 including one, known as the Bristol & English Channels Connecting Railway, a joint venture by the Bridgwater & Taunton Canal and the Chard Canal Companies, to link Stolford (near Bridgwater) with Ilminster, Crewkerne, Beaminster and Bridport, with branches to Taunton, Chard, Axminster and Lyme Regis. This foundered, as did the fortunes of its companions, and almost 60 years would pass before trains would run to the resort. However, although the GWR had been kept out of the area due to the goings on at Chard and Exmouth, there was still a danger from the broad gauge company now that in 1857 it had reached Bridport, the next sizeable town eastwards, with a possibility of it striking out to Charmouth and Lyme Regis. The LSWR was keen to

encourage proposals for a line to Lyme in the early 1870s, but this was only to keep out its old adversary and, once this had been accomplished, the South Western dropped all interest in the matter. To state that the LSWR was devoid of interest in the construction of a line to Lyme Regis would be an understatement, as it had been effective in blocking one proposal (with the assistance of the GWR) in the 1860s, and would do so twice more in the 1880s. The LSWR was, however, eventually persuaded to agree to the principle of working the line, that is if one could actually be built, with the result that the Bill for a Lyme Regis Railway Company to construct a 7¼-mile branch at a cost of under £86,000, to connect with the LSWR, was given Parliamentary approval in 1871. Plans were also submitted to extend the proposed line onwards to Bridport and yet, for all this fervour and Parliamentary success, plus a sod-cutting ceremony on 29 September 1874 performed by the Mayoress, Mrs Skinner, the plans came to naught, with the powers eventually lapsing and still no indication of when railway communication would be realised.

There were demands not only from the inhabitants of Axminster for the connection, but also the townspeople of Chard were keen to have the route provided from their area. The spectre of the Bristol & English Channels Railway, with its plan for a line to Lyme Regis via Chard Junction and Hawkchurch, was still there although the line would have shortened the journey to London and could have avoided the need to construct a further junction. The Government had passed the Light Railways Act in 1896, thus enabling branch lines to be made on the cheap, subject to certain restrictions, with basic signalling and simplified level crossing arrangements. Over 2,000 people signed a petition which was presented to the South Western in 1898 but, true to form, this was also ignored, although due to the intervention of Sir Wilfred Peek, an influential local public personage, the LSWR was finally coaxed into life with the result that the Axminster & Lyme Regis Light Railway Order dated 15 June 1899 was granted, enabling construction of a line from Axminster to a site about half a mile from the centre of Lyme Regis. At long last, the hopes and dreams of a line for the local people were now to be realised after so many setbacks

Above: Rebuilt Bulleid Pacific No 34003 *Plymouth* runs into Axminster with the 10.42am Salisbury to Plymouth on 8 July 1961 while 'U1' No 31901 (a three-cylinder development of the Maunsell 'U' class) stands on the up main with the 10.42am Exmouth to Cleethorpes. The train was formed of Eastern Region and Southern stock on alternate weeks. *Terry Gough*

OVER BRIDGE
for SALISBURY LONDON
& LYME REGIS TRAINS

Evening shadows begin to fall as 'S15' 4-6-0 No 30841 — with plenty of steam to spare for the gruelling climb over the Honiton bank — clanks into Axminster with a down freight on 12 August 1960. The headlamps are already lit and in place as it will be dark upon arrival at Exmouth Junction. *Terry Gough*

and failures. The appointed engineer of the 6¾-mile line was Mr Arthur C. Pain who had been involved with the Culm Valley line and the Southwold Railway. Standard gauge was stipulated, with an axle loading of 12 or 14 tons (according to the weight of the rail) and a maximum speed of 25mph. A tender of £36,542 from Baldry & Yerburgh of Westminster was accepted on 10 April 1900, with a total of £50,360 having been subscribed, this including £25,000 from the LSWR, whose involvement with the local company resulted in the South Western receiving 50% of the revenue. All shares had been taken up from 5 July the same year, and work started on 21 June 1900, but the contractors experienced difficulties in construction of the line and an extension of the Light Railway Order was needed.

Much of the heavier materials used, including ballast, sleepers and rails, were brought in by boat, not surprisingly considering the narrow lanes of East Devon. The first such cargo was unloaded at the Cobb in Lyme Regis in early August 1900. The English Channel, well known for its westerly gales, often disrupted deliveries. Other items brought in by sea

included equipment for the impressive Cannington Viaduct which, at the time of building, was the second highest of its type, being one of the earliest examples of concrete construction. Built on a grand scale by 'Concrete Bob' McAlpine, fresh from similar engineering constructions on the Fort William to Mallaig line, the viaduct itself consisted of 10 elliptical arches each of 50ft, the total length being 600ft on a gradient of 1 in 80 and the height from the valley floor to rail level being 92ft. Two wooden pylons supported a 1,000-span cableway upon which materials and concrete were conveyed across the valley for the construction works. The concrete was hand mixed using crushed flint removed from the chalk excavated from nearby cuttings. Difficulties with pier number one and the west abutment, which settled vertically due to the greensand and blue lias of the valley floor not accepting the design load of 3 tons per sq ft, resulted in a 'Jack arch' being built in the third span. The formation of the trackwork was built up slightly to compensate, giving an unusual switchback feel to this end of the viaduct. As a result of the settlement, a flagman was posted at each end of

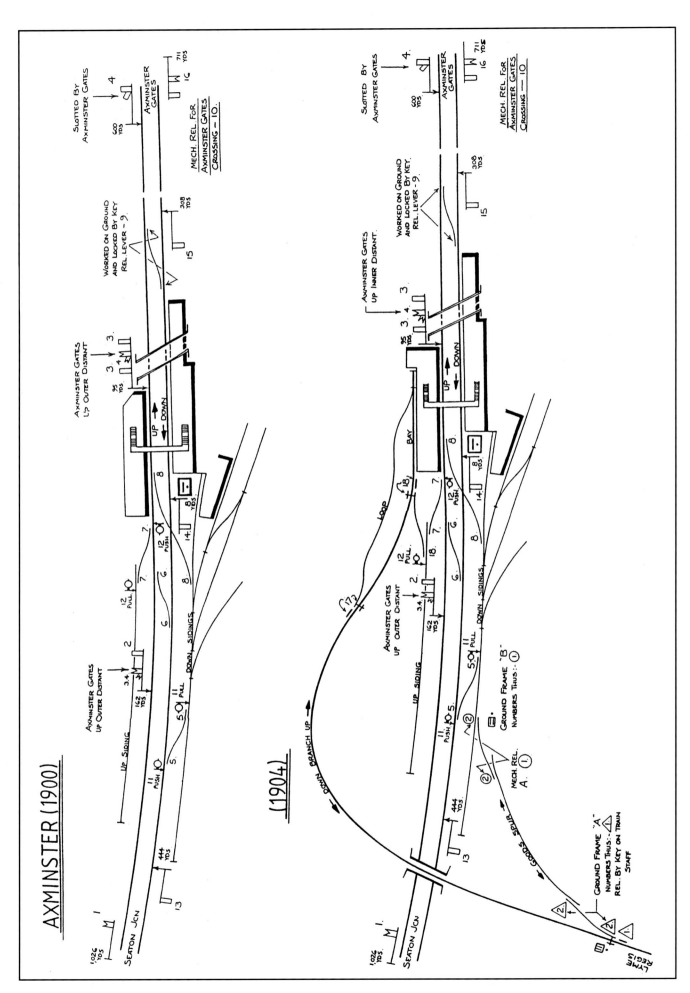

AXMINSTER (1900)

(1904)

Adams Radial '0415' class 4-4-2T No 30582 runs round its train in the branch bay at Axminster on 16 August 1959. The water tower, so long a feature of the station, is visible in the background. The signalbox and down platform starter can be seen to the right, with the Sykes banner repeater for the up starter situated on the up platform to the right of the coach. *Terry Gough*

Above: Change-over day at Axminster on 25 May 1935. No 3125 stands in the bay on the branch set and No 3520 waits alongside the up platform for the road to Exmouth Junction shed. A down express headed by 'N15' No 455 *Sir Launcelot* stands on the down main line. *H. C. Casserley*

the viaduct to keep watch for any untoward signs of instability but, such was the quality of the construction, there was never any sign of movement reported during the 62 years of the line's operation. The South Western withdrew the flagmen when it gained control of the line, with a speed limit of 15mph being adhered to on the viaduct.

Axminster

Axminster, with its large impressive station buildings with their tall chimneys and pitched gables (the hallmark of Sir William Tite) dominating the down platform, opened with the main line in July 1860. The station was well patronised, being better served by passenger services than its neighbour at Chard Junction. Expansion and growth came with the doubling of the line by July 1870, the Waterloo-Exeter trains in the 1860s being worked by the Beattie designed 2-4-0s. The early locomotives allocated to Exeter were supplied with spark arrestors and No 71 *Alaric*, a 2-4-0 of the 'Falcon' class, departed with

such force from Axminster on 19 April 1872 that the spark arrestor was forced from the chimney by the blast. Unfortunately it landed under the nose of a horse, which subsequently panicked and came to grief when it tried to jump a hedge while still attached to a carriage. This incident cost the LSWR the sum of £25 to keep out of court, and the Exeter locomotive foreman was instructed to remove all spark arrestors without delay.

Considerable alterations to the layout at Axminster were necessary for the new branch which crossed the LSWR main line by a girder bridge, curving and running down a 1 in 80 incline to a bay line behind the up main line platform. A run-round loop was provided, and a coaling stage and water column for the branch engines. The water tank, on its tall brick-built base, complete with a brick chimney for the stationary steam engine used for pumping water from the River Axe, was a well-known feature at this end of the station until the chimney was removed in early 1961. There was a direct connection from the branch to the down goods yard, operated by two ground frames. The earthworks for the branch required the

Above: No 30582 runs into the bay platform at Axminster with the 9am Lyme Regis to Waterloo on 13 August 1960. No 30584, which had double-headed from Lyme Regis and had cut off upon arrival at Axminster, can be seen in the left background. *Terry Gough*

demolition of a small corrugated iron locomotive shed, dating from the 1860s, which was situated at the Exeter end of the yard. Although, on paper, the new goods link was a useful feature, offering branch freight trains direct access to the yard without disrupting traffic on the main line, as is often the case it was found to be difficult to work as down freight trains for the branch were faced with stopping twice for the ground frames to be worked and having to restart on a 1-in-40 gradient. This proved to be no easy matter and the goods spur was removed from 5 September 1915, after which branch freight trains shunted across the main line via the bay and up siding.

When the line opened it was operated under the 'one engine in steam' regulations and, as no signals were provided, only three extra levers were installed in Axminster signalbox. Alterations took place in 1906 to the branch signalling, with the Axminster box frame being extended to 30 levers plus the addition of the Tyers No 6 tablet instrument. The branch signalling at the main line station consisted of an up fixed distant signal (1,054yd), an up inner home and a platform starter. Retraction came from 27 March 1960

when the branch signals at Axminster, except for an up fixed distant, were removed, the Tyers No 6 tablet being replaced by 'one train' working, thus reverting to the practice at the opening of the line. The run-round points for the branch trains were disconnected from the signalbox and operated by new ground frames released from a key on the train staff. The signalbox at Lyme Regis was reduced to a ground frame at the same time and closed entirely on 20 July 1965, with all lines removed except for the platform line.

Axminster was well served by a selection of semi-fast and local stopping services, including the cross-country Brighton-Plymouth services. Through coaches to and from Waterloo appeared in the summer timetables from June 1953 to September 1963, and until 1959, there was one return service on Mondays and Fridays with two on Saturdays. Thereafter they ran on Saturdays only, with an unbalanced two up and one down service. The 10.45am from Waterloo carried a rear portion of five coaches bound for Lyme; the Radial tanks would then make a double-headed trip, albeit without the branch

Above: The through coaches have been left at Axminster by the 10.45am from Waterloo. They have been collected by No 30584 seen here passing the signalbox en route to the up siding, and then reversed into the branch bay. The assisting locomotive will then couple up for the journey to Lyme Regis. *Adrian Vaughan collection*

Below: The change-over Radial tank, No 30584, arrives at Axminster from Exmouth Junction on Saturday, 13 August 1960. *Terry Gough*

The 1 in 80 gradient from the station is tackled with ease by No 30583 leaving Axminster with the 12.6pm to Lyme Regis on 6 September 1953. *S. C. Nash*

Above: Nos 30583 and 30582 arrive at Axminster on 27 June 1959 with the 9am Lyme Regis to Waterloo. *S. C. Nash*

Below: Having used the run-round loop in the branch bay at Axminster, No 30582 reverses on to its single coach for the return trip to Lyme Regis on 25 May 1956. *Adrian Vaughan collection*

Above: A sharp November frost decorates the main line sleepers at Axminster in 1958 as ex-GWR 0-4-2T No 1462 stands on the up main awaiting trials on the Lyme Regis branch. Unfortunately it was going to be a very bad day for this particular locomotive. *John Day*

Below: 'M7' No 30048 has arrived at Axminster from Seaton Junction with the Seaton branch set (No 603) on 1 October 1960. Single-line working was in operation due to a bank slip on Honiton Incline. *A. E. West*

Above: Bulleid 'West Country' class 4-6-2 No 34036 *Westward Ho* arrives at Axminster with a Waterloo to Exeter service on 12 March 1952. *A. E. West*

Below: Adams Radial tank No 30584 heads for Axminster at Trill with an up train on 30 August 1953. *S. C. Nash*

Above: This is what made the Lyme Regis branch such a special line for travellers and enthusiasts alike. Two Adams Radials, Nos 30582 and 30583, at Trill with the 10.45am Waterloo to Lyme Regis on 9 September 1960. *S. C. Nash*

set, and at the terminus the staff had about 40 minutes to clean and prepare the stock as the train was booked away at 3.5pm for Waterloo, being attached at Axminster to the 12.45pm from Torrington. Branch services also connected at Axminster from the summer of 1960, with the 7am Cleethorpes to Sidmouth and Exmouth, and also the 10.45am Exmouth and Sidmouth to Cleethorpes. The 1pm from Waterloo detached a coach for Lyme at Templecombe which was brought down by a local stopper and left alongside the down platform for the branch engine to collect. An up through coach from Lyme was attached to the rear of the 8.55am ex-Ilfracombe which, upon arrival at Salisbury (with the through Yeovil Town coach), was attached to the up 'Atlantic Coast Express'. Passenger traffic was very heavy during the summer months and even on Sundays there was no respite, with excursion traffic arriving from Taunton via Chard, and from Bath via the Somerset & Dorset. With the exception of the Brighton-Plymouth service, through trains to Plymouth from Waterloo ceased in September 1964 and from then on it was to be a period of dereliction along the whole route, with the Lyme Regis branch closing on 29 November 1965. Axminster signalbox was abolished on 5 March 1967, and the main line singled from 11 June the same year. All trains today use the former down platform, with the station having a good patronage, especially for travel to Exeter.

Combpyne

There was only one intermediate station, at Combpyne 500ft above sea level, situated ¾ mile from the village bearing the same name, and located 4¾ miles from Axminster and 2½ miles from Lyme Regis. The 250ft bare, single platform, lit at night by Tilley lamps, was unusual in that the main brick-built station buildings containing the ticket office, waiting room and ladies' and gents' toilets, plus the stationmaster's house, were remote from, and at right angles to, the platform. The first stationmaster was Mr Greenslade, who for many years previously had been a guard on the LSWR. Signalling came to the line from July 1906 with signalboxes being opened at Lyme Regis and Combpyne. The box at Combpyne was positioned off the down end of the single platform, and a passing loop was provided at the same time, the signalling consisting of fixed distant signals, positioned at 1,015yd (up) and 1,006yd (down) respectively, plus an outer home and advanced starter for the up loop, and an inner home and starter protecting the down loop. A siding, which in itself was a loop off the up loop with two short sidings, was also provided and as at Lyme Regis the box was equipped with the Tyers No 6 tablet instrument. Seeing little use as a block post except during the seasonal traffic, the box was closed from 12 August 1921, all instruments and

Above: A view of Combpyne taken post-1921 showing the original signalbox. The box was closed from 12 August that year with all instruments and signalling removed and reduced to a ground frame for the truncated sidings until it was closed and was replaced by a two-lever ground frame on 17 June 1930. The box was then removed to nearby Hook Farm. An unidentified Adams Radial tank heads a train to Lyme Regis. *Lens of Sutton*

Right: A pair of Adams Radial tanks working in tandem was always one of the pleasures of the Lyme branch. Nos 30584 and 30583 near Combpyne with the 4.36pm Axminster-Lyme Regis on 6 June 1960. *S. C. Nash*

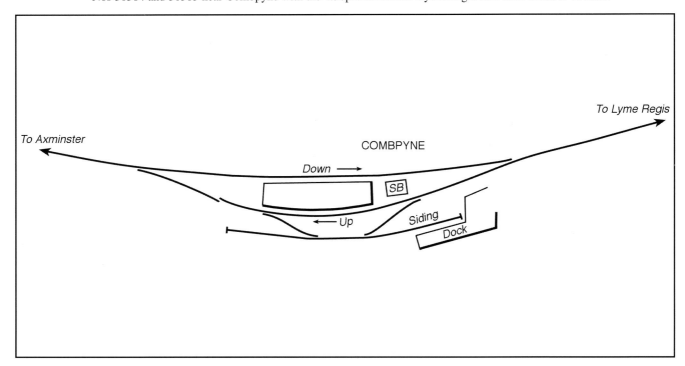

Above: An almost timeless branch line scenario — a locomotive with a single coach on a fine summer's day. No 30582 trundles towards Axminster through the cutting at Combpyne on 14 June 1949. *S. C. Nash*

signalling being removed. The passing loop was taken out of use at the same time, and with the former down loop then being used as the single line and the up loop reduced to siding status, the box became nothing more than a ground frame for the truncated sidings until it was closed and replaced by a two-lever ground frame on 17 June 1930. The frame was unlocked by the key on the end of the single-line token. The former cabin was removed to nearby Hook Farm and used for storing grain. Other alterations at the time included the complete removal of the siding formed out of the up loop, leaving the remaining siding, which was now a dead end, connected to the Lyme Regis end of the single line. A new platform face was built from concrete slabs, while the other face was formed into a gently sloping grassed bank. A boost to traffic occurred in 1908 when a landslip took place in the

cliffs south of Combpyne, with the ground catching fire and burning spectacularly for eight months, the branch trains bringing many spectators to view this event. Later years witnessed the appearance of an elderly ex-LCDR carriage which was provided as a camping coach during the late 1940s and proved popular with summer visitors. As Lyme Regis was famous for old fossils, this specimen of coaching stock fitted the bill admirably! However, a larger camping coach in the form of an ex-LSWR composite was provided later. Campers were a hardy breed; the coaches had no heating or running water, in fact the water was supplied from Lyme Regis in milk churns, with the station staff at Combpyne being responsible for issuing new blankets etc. The camping coaches were returned to Eastleigh Works for maintenance in the winter.

Above: A member of Combpyne station staff carries empty water cans to be returned to Lyme Regis for refilling as No 30584 arrives with the 12.30pm from Axminster on 7 July 1959. *H. C. Casserley*

Below: No 30584 leaves Combpyne for Axminster with coach No S6595S on 6 July 1959. A water can has been left on the platform and clothes dry in the summer sunshine to the left of camping coach No S38S. *H. C. Casserley*

Above: Camping coach No S38S at Combpyne on 18 September 1961. This vehicle, converted from SR composite No 5056 in 1954, was built at Eastleigh in 1906 and went for scrap in 1968. The vehicle is in green livery with yellow numbers and 'Camping Coach' in white. *A. E. West*

Below: Combpyne, looking towards Lyme Regis, on 15 July 1958 with the station buildings seen to the right. Note the curving trackwork of the branch line. *H. C. Casserley*

Above: Ivatt Class 2MT 2-6-2 No 41297 was tested on the branch on 18 September 1960 and is pictured here in the yard at Combpyne. Local children gather to see what all the fuss is about as the civil engineer and his staff (complete with gauging bar) check the trackwork. *S. C. Nash*

Below: Heading for Lyme Regis, the LCGB 'East Devon' rail tour steams through Combpyne on 28 February 1965 hauled by Ivatt 2MT 2-6-2T No 41291 with No 41206 assisting at the rear. *Hugh Ballantyne*

Above: The LCGB 'East Devon' rail tour returning from Lyme Regis traverses Cannington Viaduct 'top and tailed' by Ivatt 2-6-2Ts Nos 41206 and 41291 on 28 February 1965. *Hugh Ballantyne*

Lyme Regis

A '330' class 0-6-0 saddle tank, No 131, was offered to the contractors by the LSWR and was at work on the line in 1902-3. The locomotive, built by Beyer Peacock & Co, was one of six delivered in May 1876 at a cost of £2,375 each. No 131 was at first allocated to Nine Elms and was sent to Northam in 1881; by March 1883 it was allocated to Basingstoke, and was at Bournemouth in March 1912 with the duplicate number of 0131, eventually being withdrawn and scrapped in 1924-5. The locomotive hauled the inspection train on 22 January 1903 conveying directors of the Axminster & Lyme Regis Light Railway, together with engineers from Devon and Dorset, but it was found to be unsatisfactory for working service trains and was withdrawn. The actual opening of the line was delayed due to some movement in earthworks, and a further inspection by Major Druitt, acting on behalf of the Board of Trade, took place on Friday, 21 August. Approval was sanctioned and the 6¾-mile line, of which only the final half a mile was actually laid in Dorset, opened three days later on Monday, 24 August 1903 amidst

the usual celebrations and euphoria, with the first train departing from Lyme Regis in the rain at 9.40am. Amongst the passengers were 200 children from the town having a 'free ride', the cost of their tickets being met by public subscription — a nice gesture. Unfortunately, no subscriptions were raised in Axminster for their scholars. The single fare at the opening was 6½d. The main train, headed by the two former LBSCR 'Terrier' 0-6-0 tanks, Nos 734 and 735 suitably decorated with garlands and flags, consisted of 13 four-wheeled coaches and left Lyme Regis at 12.25pm in perfect sunshine, conveying various worthies and dignitaries. All this was watched by a large crowd of onlookers enjoying the proceedings, and when the inaugural train arrived at Axminster the Pride of the Axe brass band struck up with 'See the Conquering Hero Comes' followed by 'Rule Britannia' and the National Anthem. A champagne buffet was held in the waiting room on the down platform and was enjoyed by the official party, who then returned on the 1.18pm train for a sumptuous lunch at the Red Lion Hotel in Lyme Regis.

The whole branch was subject to a maximum speed of 25mph, with a restriction of 15mph over

Above: The stationmaster and staff, plus plenty of onlookers, stand proudly beside the two ex-LBSCR 'Terriers', Nos 734 and 735, bedecked with flags and garlands at Lyme Regis on the opening day, 24 August 1903. The leading engine, No 734, has been preserved and can be found working on the Isle of Wight Steam Railway. *Lens of Sutton*

Below: An early picture of Lyme Regis station shows a rear view of one of the four-wheeled coaches alongside the platform. Part of the goods shed can be seen in the background before being moved in 1904 to a position nearer the Axminster end of the layout. *Lens of Sutton*

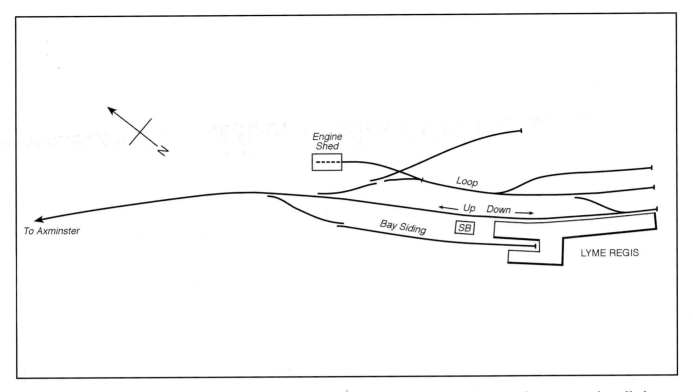

Cannington Viaduct and around curves of less than 9 chains radius speed was reduced even further to 10mph. The line was built to light railway standards which meant lightweight 56lb flat-bottomed rails flanged and bolted to the sleepers, sharp curvatures and severe gradients. The latter two features were to haunt the operating and civil engineering departments of the LSWR, Southern Railway, and British Railways almost to the end of steam operation itself. The line cost £11,800 per mile to construct including purchase of the land. At first there were no fixed signals, the line being worked under train staff regulations with telephone communication used to inform stations when trains were despatched. The line was operated entirely by the LSWR under contract to the Axminster & Lyme Regis Light Railway. The original coaches were a small composite, with the words 'first' and 'second' obliterated, and two small brake thirds, all being four-wheeled vehicles which were replaced with steam-heated bogie stock by the LSWR. Due to the curves on the line, six-wheeled stock was prohibited, unless 'absolutely necessary', as the LSWR's 1911 Appendix Instructions put it.

The station buildings at Lyme Regis were of timber construction, the station being situated 249ft above sea level at the end of a steep climb from the town. The layout consisted of a single platform measuring 300ft long by 14ft wide served by a run-round loop and a bay siding. Opposite the station there were three sidings, a timber-built goods shed near the yard entrance (which by 1904 had been resited to a position opposite the Axminster end of the platform) and a W. H. Smith bookstall which was originally placed at the northern end of the station. This was used later as a store, a new bookstall being erected at the busier, south end of the station but later removed to St Mary

Cray in Kent. A 15cwt yard crane was installed near the northern end of the goods shed. After closure of the branch this goods shed was destroyed by fire in December 1966. In 1904 the station was rebuilt with sidings enlarged and altered, and in Southern Railway days the station building was lengthened and the booking office transferred from the middle portion and reinstated at the end nearest the town. A pleasing ornamental roof awning complemented the design of the station. The first stationmaster was Mr Ley who had started his railway career as an office lad in the goods department at Poole in 1882, moving to Hamworthy upon promotion in 1885 and remaining there for 10 years before relocating to Poole as booking clerk for eight years. Promotion took him to Lyme Regis in 1903.

Annual passenger traffic by 1910 had reached a total of 60,000, with 19,000 parcels and a coal and goods tonnage of just over 9,000 tons. The ground level box at Lyme Regis stood off the end of the platform at the Axminster end and was equipped with 14 levers and a Tyers No 6 tablet instrument. Signals consisted of a fixed down distant located 1,016yd out from the box, a down home, a platform starter and an up advanced signal, plus associated ground signals for the pointwork etc. Prior to the opening of the box, the run-round loop was controlled by a five-lever ground frame operated by the key on the wooden train staff. The tablet instrument was moved to the parcels office from 1 June 1930 to permit the signalman to assist with station duties, he returning to the box when needed to operate the signalling apparatus. The wooden name board fixed to the front of the signalbox was removed and replaced by an enamelled name plate on the north-west end of the box where it could be seen more easily by passengers looking out of the windows of trains running into the station. Station

Above: One of the 'Terriers' awaits the 'right away' to leave for Axminster with a train of four-wheelers; a road van used for loading and unloading of parcels is coupled next to the locomotive. The goods shed has yet to be resited. *Lens of Sutton*

staff consisted of a stationmaster, two signalmen, one goods porter and a van driver.

At the opening of the line, six return journeys were provided on weekdays only, with the first train departing from Lyme Regis at 9.40am and the last returning at 9.15pm, the running time for the 6¾ miles being 21 minutes. The LSWR gradually increased the services upon absorption of the line in 1907 and the winter timetable for 1907-8 shows eight passenger trips and one goods trains each way daily. Five or six of the passenger trains worked as mixed trains. World War 1 brought a severe reduction in passenger traffic, but this picked up at the end of the war. Through traffic continued to be heavy due to the many excursions arriving from London, but by the 1920s road competition was beginning to come to the fore. In February of that year, motor omnibuses began working a Bridport-Charmouth-Axminster service and by the summer of 1922 they were running between Axminster and Lyme Regis. The awkward location of the station at Lyme Regis with its steep climb from the town only added to the attraction of the motor bus for local people.

The Southern Railway introduced regular summer Sunday services in 1930 for the first time, with return train journeys ranging from six to 11 trips. This was extended throughout the winter in 1940, but reverted to summer Sundays only in 1951. The winter Sunday timetable displayed the local bus services, which omitted Combpyne. Weekday summer seasonal services reached their peak under the Southern

Railway prior to World War 2 with 11 return journeys, and except for a brief interval after the end of the war, the services never reached such a figure again. Wartime trains consisted of a sparse regular service throughout the week, but from 1951 until closure of the line, a total of nine weekday return journeys per day were made with extras on Saturdays, with the summer Sunday service ranging from eight to 11 return trips. Amongst regular travellers who returned to the line in June from year to year were the Oldham Boys Brigade travelling to their annual camp at Lyme Regis in a six-coach train formed of London Midland Region stock. The coaches were kept in a siding at Lyme for the duration of the weekly camp. The RCTS ran a special train to Lyme Regis on 28 June 1953 to celebrate their 25th anniversary and this was headed from Axminster by Adams Radial tank No 30583, now preserved on the Bluebell Railway, and ex-LBSCR 'Terrier' No 32662, which is now preserved at Bressingham, Norfolk.

The two Saturday services from Lyme Regis which had through coaches to Waterloo in the summer months of 1958 were the 9am and 3.5pm departures. When the Saturday relief Adams tank arrived at Axminster from Exmouth Junction it would be attached to the 8.32am service and double-head to Lyme. Upon arrival at the terminus the following ritual would take place: on stopping at the home signal, the pilot engine would detach and pass into the goods yard, allowing the train engine and its single coach to enter the bay road. The pilot engine would

Above: Private owner coal wagons stand in the yard at Lyme Regis including one (No 134) belonging to the well-known firm of Bradford & Sons Ltd, Yeovil. The bookstall of W. H. Smith can be seen at the far end of the platform.
Lens of Sutton

then reverse out of the goods yard and attach to the coach in the bay road, reverse out therefrom and propel the coach on to the four coaches for Waterloo standing on the platform line. The train engine would then reverse out of the bay road and attach to the locomotive standing on the stock on the platform line thus forming the double-headed 9am departure to Waterloo. Only seven minutes were allowed for this operation, and upon arrival at Axminster the four coaches would be shunted and attached to the 8.30am Exeter Central-Waterloo at 9.24am. During the 1950s, a total of 8,000 passengers were carried on the branch during the summer period compared with a maximum of 1,500 during the winter months. However, the permanent withdrawal of the through Waterloo service at the conclusion of the 1963 summer timetable affected the branch services. Despite this, the passenger traffic remained healthy, but all to no avail as the last trains ran on Saturday, 27 November 1965 with a three-car DMU working the services. The platforms at Axminster and Lyme were packed with intending passengers all day long — far too many for the train to convey — and when the 3.39pm departed from Lyme Regis, a total of 320 passengers were on board, leaving about 60 on the platform for whom the train made a special return trip to collect. The final

train departed from Lyme Regis at 7.10pm with arrival at Axminster to the traditional barrage of detonators. The DMU relinquished the single-line token for the last time, was released on to the main line and departed into the darkness for Exeter.

The line to Lyme Regis, so much fought for in the earliest years from its conception through to opening, was now extinct. Two of the locomotives which worked on the line regularly still survive, and a part of the station building at Lyme Regis was removed and has been rebuilt at Alresford on the Mid-Hants Railway where it is in use as a gift and book shop. The track was lifted from the branch completely by 28 June 1967. Plans by Minirail to provide a 15in gauge railway on the trackbed of the former line with headquarters at Combpyne unfortunately did not come to fruition, although a 1½ mile of track was actually laid and locomotives and stock gathered. A small industrial estate formed of small units covers the site of Lyme Regis station today, and nothing exists to remind visitors of its railway history. At Combpyne, the former station house has been converted into a private residence and the platform removed, the area it once occupied having been made into a lawned area, although the small loading dock still survives.

Above: On a typical wet summer's day, umbrellas and raincoats are essential. No 0520's train has discharged its passengers after arrival from Axminster on 26 August 1928. *H. C. Casserley*

Below: The locomotive shed doors stand closed as No 3520 (former LSWR No 520) awaits the next trip to Axminster on 31 August 1945. Renumbered No 30584 in BR days, this locomotive dating from 1885 was scrapped in 1961. *H. C. Casserley*

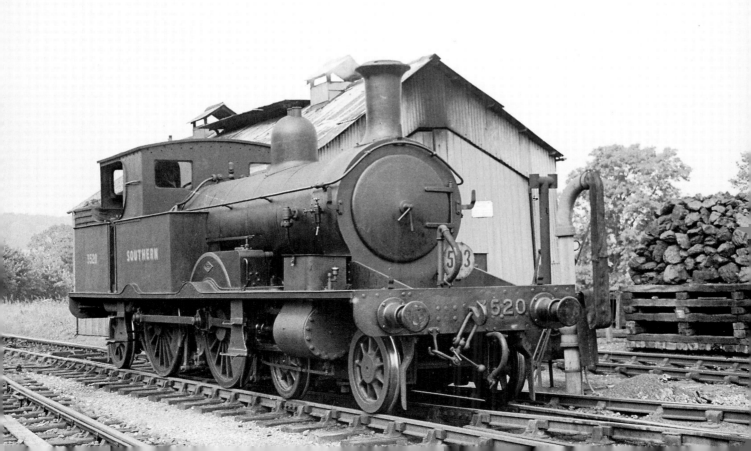

Above: Lyme Regis as viewed from the station frontage on 7 July 1959. The coach standing adjacent to the stop block is No S970S, an ex-SECR 60ft 1in third built at Ashford in June 1923 and withdrawn in November 1959.
R. M. Casserley

Below: A hot summer's day at Lyme Regis and No 30584 runs round the stock after arrival from Axminster on 7 July 1959. The pleasing station building adds to the attractive scene. Coal wagons stand in the yard on the right and the goods shed can be seen in the background. *H. C. Casserley*

Above: Ivatt 2-6-2T No 41322 awaits its next turn of duty while standing near the single-road locomotive shed at Lyme Regis. *John Day*

Below: Box vans stand alongside the goods shed at Lyme Regis as viewed from the locomotive shed road on 12 July 1962. The signalbox and station bask in the summer sun during a traffic interlude. *John Day*

Above: An ex-LSWR 20-ton goods brake, No S54977, stands in the yard at Lyme Regis on 17 July 1955. The side doors were for ease of loading and unloading parcels at stations when in use as a road van. The goods brake is in grey livery with a buff interior. *A. E. West*

Lyme Regis Locomotive Shed

A single-road locomotive shed measuring 100ft x 20ft was built to house the tank engines. This building was constructed of timber and there was a sleeper-built coal stage outside the building with a water column positioned near the shed entrance. No suitable LSWR locomotives could be found for the opening of the lightly laid line; the '330' class tanks originally intended to be used on the services proved unsuitable following the problems with No 131 which made it necessary for other motive power to be used. Dugald Drummond, who had initially considered constructing two small 0-4-4 tank engines for the purpose, approached Robert Billinton, the Locomotive, Carriage & Wagon Superintendent of the LBSCR at Brighton Works, regarding the purchase by the LSWR of two Stroudley 'A1' class 0-6-0 'Terrier' tanks, as mentioned previously. After negotiations, and with the approval of the LSWR Locomotive Committee, Nos 646 *Newington* and 668 *Clapham* were duly delivered in LBSCR livery to Nine Elms on 14 March 1903 at a cost of £500 each. The South Western certainly had a bargain; the engines were taken into the Works to have their Westinghouse air-brake equipment removed and replaced by vacuum ejectors.

They were painted in LSWR livery, the LBSCR number plates removed and the locomotives renumbered as 734 and 735 respectively. It is rumoured that Drummond used the cast brass number plates on his desk as paperweights. (He had been Works Manager at Brighton under Stroudley from 1870 to 1875.)

Entry into South Western service of these two locomotives was swift, with No 734 being despatched to Guildford, working as shed pilot for several weeks, before being sent to Exmouth Junction where duties included Exeter Queen Street carriage pilot and working the Exmouth, Budleigh Salterton and Sidmouth goods trips. The locomotive is reported as being derailed in the sidings at Topsham on 12 June 1903. Meanwhile her stablemate, No 735, had been put to work around the shed yard at Nine Elms for a few months before being transferred to Yeovil for the Town-Junction shuttles. Both locomotives are recorded as being at Exmouth Junction shed on 17 July 1903, with No 735 running a series of trials over the Lyme Regis branch in early August 1903 while No 734 was employed on the Sidmouth Junction-Sidmouth services on August Bank Holiday Monday. It was found that during the winter months a single engine could work the passenger and goods traffic, being sub-shedded at Lyme Regis and changing over weekly with the spare engine from Exmouth Junction on Monday mornings before the

Below: Ivatt '2MT' class 2-6-2T No 41322 stands alongside the platform at Lyme Regis on 12 July 1962. Built at Derby in February 1952, the locomotive was withdrawn in June 1964. Thirty-five members of this class of locomotive were at one time working on the Southern Region. *John Day*

start of passenger services. The spare 'Terrier' at Exmouth Junction was employed on local duties as described above. However, many of the Lyme services in the summer required double-heading which, as well as being uneconomical, could prove to be a headache for the motive power department especially if one of the 'Terriers' was stopped for repair at Exmouth Junction. To make matters worse, the civil engineer had prohibited a 'Terrier' and 'O2' pairing, and working hard on the severe curves and gradients was causing problems for the ex-LBSCR tanks which had to run with tanks no more than half full in order not to exceed the axle weight of 12 tons.

The 'Terriers' were in sole charge of the line until August 1905 when the first of the Adams 'O2' 0-4-4 tanks arrived to give assistance. Unfortunately, due to the restrictions of the line, these locomotives, although stronger machines than the 'Terriers', were unable to give their customary performance. So it was

'back to the drawing board' for the operating department, leaving Nos 734 and 735 to continue working the services to and from Lyme for the time being. 'O2' No 228 was sent to Lyme Regis in January 1906 and, following further trials, when its side tanks were marked with the maximum water capacity to keep within the restrictions of the line, it began work on the branch that summer. The line was not the immediate financial success that the Axminster & Lyme Regis Railway had envisaged, with the result that the LSWR absorbed the company on 1 January 1907. It then started to make improvements to the branch, with the engineers department relaying much of the track with heavier rails, albeit second-hand, easing the sharpest of the severe curves, adding drainage works and building a short connecting spur to the down platform at Axminster. However, it was 1912 before all the original flat-bottomed rail had been replaced.

Services were taken over by specially modified Adams 'O2s' Nos 177, 191, 202 and 228 from May 1907 and mid-1909 found the 'O2s' working in pairs on most summer Saturdays. The 0-4-4Ts, although a satisfactory locomotive on other parts of the South Western, began to suffer the usual problems of distorted frames and excessive flange wear due to the excesses of the route.

The locomotive shed was burnt to the ground on 28 December 1912 and it was only due to the prompt action of the railway staff that the locomotive inside was saved from serious damage. Plans were approved for a new building in April 1913, at an estimated cost of £400. The new shed, only about half of the length of its predecessor, was constructed on the foundations of the original, with a steel frame clad in corrugated iron and had wooden doors and steel window frames. An inspection pit ran the length of the building, complete with washout hydrant, and two raised smoke vents in asbestos were provided on an asbestos roof. Much of the locomotive coaling was done at Axminster. Footplate staff consisted of two drivers, two firemen and an engine cleaner. The shed closed on 4 November 1963 with the withdrawal of steam locomotives from the line.

Robert Urie, who had been appointed Chief Mechanical Engineer for the LSWR upon the death of Dugald Drummond in 1912, took an Adams '0415' class 4-4-2 Radial tank, No 0125, dating from 1885, into the workshops at Eastleigh in September 1913 as an experiment, modifying the bogie to give greater side play in order to cope with the severe curves which had plagued the 'Terriers' and the 'O2s'. The modified locomotive departed from the Works on 11 October and was tested five days later on both goods and passenger trains. The alterations made the Adams tank a success, and it was ideally suited to the line with its 5ft 7in driving wheels and short-coupled wheelbase combined with a light weight of 54 tons spread over five axles. The civil engineer was also satisfied with the locomotive, apart from recommending that the quantities of coal and water should be reduced, and it was put to work on the branch services in November 1913. Two more Radial tanks were similarly modified early in the new year — No 521 (24 January 1914) and No 0419 on 7 March.

Normally, one locomotive was kept at Lyme Regis while the other two were based at Exmouth Junction and used mainly on the Exmouth services. As before, in busy periods two locomotives were used on the Lyme Regis branch, but with three Radials now on hand it was possible to cover for routine repair work without using the 'O2s', although the 0-4-4 tanks were still used on occasion. The '0415' class in service throughout the system had shrunk to a total of three by 1 January 1927 with Nos 0125, 0486 and 0520 being reserved for the Lyme Regis branch. However, No 0486 was withdrawn in January 1928 leaving the two survivors facing what was thought at the time to be an uncertain future. Searching for a replacement for the Radials brought an ex-SECR 'P'

class 0-6-0T, No A558, and ex-LBSCR 'D1' class 0-4-2 tank No B612 (formerly LBSCR No 12 *Wallington*) to Exmouth Junction in October 1928 for trials on the line. The 'P' class tank was found to be totally inadequate and no improvement on the 'Terriers,' but the 'D1' was more successful, although the locomotive was too heavy and had to be modified at Brighton Works. During March-September 1929 Nos B276 (*Rudgwick*), B359 (*Egmont*), B612 and B633 (*Mitcham*) had their bunkers cut down to reduce coal capacity, and by restricting the water tank to 580 gallons a weight saving of more than two tons resulted. The four engines were then despatched to Exmouth Junction for use on the Lyme branch.

The two Radial tanks, Nos 0125 and 0520, were put aside awaiting the call to Eastleigh for withdrawal and breaking up but, by a stroke of good fortune, the Works at that time was snowed under with work, and the Exmouth Junction shedmaster had the foresight to have the pair tallowed down and moved into the shed. This precaution proved to be fortuitous, as the 'D1s' were less suited to the severe curves than had been anticipated. By April 1930, Nos B276 and B359 were unserviceable with strained framing and leaking tanks, with No B633 back at Eastleigh for repair, leaving the remaining member of the quartet, No 612, to work the branch with various 'O2s'. The two Adams Radials which had been awaiting the call for withdrawal suddenly found themselves in the limelight and were sent to Eastleigh for rebuilding with new frames, cylinders, fireboxes, springing, tyres and well tanks. No 0125 went into the Works on 17 March 1930 returning to traffic on 24 July the same year at a total cost of £1,290, while No 0520 was at the Works from 17 April 1930 until 14 August and cost £1,330. The two locomotives had attained a mileage of 1,302,561 and 1,341,838 respectively by this date, but their return to the branch services was greeted with much relief by the crews who had been struggling with No B612 and the 'O2s'.

The two Radials were renumbered; No 0125 became 3125 in November 1933 and 0520 became 3520 in January the following year. A Drummond boiler was fitted to No 3125 in June 1936, and livery changes occurred with No 3520 being painted in unlined Maunsell green with Bulleid lettering and numerals in January 1940, and in March 1945, it was repainted in plain black, again, with Bulleid lettering and numerals. No 3125 was similarly painted and lettered in November 1941. 'O2' tanks occasionally relieved the Radials during World War 2 and immediately after, but with the cessation of hostilities it was realised that a third '0415' was required. Fortunately, one other example still existed; old LSWR No 488 had been sold to the Ministry of Munitions in September 1917, later being purchased by the East Kent Railway where it was lying derelict but intact. The locomotive was purchased for £800 in March 1946 and despatched to Eastleigh where a further £1,633 was spent on returning it to traffic with a spare Drummond boiler. It was painted in plain

Above: Fireman Granville Morgan stands alongside No 41322 at Lyme Regis on 13 July 1962 before departing with the 2.16pm to Axminster. This class of locomotive enabled double-heading to be eliminated on the branch. *John Day*

black livery and numbered 3488, this being the number which would have been carried from 1931 if the sale in 1917 had not taken place. The locomotive retained its Adams pattern single slide bars when it returned to former South Western metals on 30 November 1946. Mileage at the time stood at 1,302,761, of which 172,165 had been worked in Government and East Kent service. All three locomotives entered British Railways stock at Nationalisation and were renumbered into the 30,000 series and painted lined black, the renumbering being: 3125 — 30582, 3488 — 30583 and 3520 — 30584. It is interesting to note that the three locomotives, all dating from 1885, were manufactured by different builders, namely Robert Stephenson (3125), Neilson (3488), and Dübs (3520).

When stationed at Lyme shed, the locomotive was kept there overnight, the firebox being relit by the permanent cleaner in the early hours of the morning. With bunker and tanks replenished, the engine would be ready for service by 6am. After booking on, the early crew would depart at 6.15am for Axminster, returning with the 7.35am goods and, upon arrival at

Lyme, the wagon brakes would be pinned down, the locomotive would be uncoupled and run into the bay road and coupled on to the coaching stock. The wagon brakes would be released allowing them to run by gravity into the goods yard. The locomotive and coaching stock would then be released from the bay road and shunted into the goods yard, picking up wagons for Axminster, plus the brake van, and then the whole assemblage moved out of the goods yard and propelled into the platform line to form the 8.11am mixed train to Axminster. Amongst the cargo were parcels and churns full of water for Combpyne. The mixed train ran during the winter timetable for 1957-8, and during the summer timetable for 1958 the wagons were attached to the first or second passenger train out of Lyme. Guards were not required on the branch trains from October 1957, provided the trains did not consist of more than three coaches. Freight traffic on the branch, although heavy in its earlier years, was never demanding, consisting mainly of coal, fertiliser, building materials including bagged cement, and general merchandise in vans or sheeted wagons inwards to Lyme. There was a weekly fitted

Above: Ivatt 2-6-2T No 41322 runs past the local platelayers' hut and into Lyme Regis with the 8.43am from Axminster on 13 July 1962. *John Day*

van for Boots the Chemist, which arrived at Lyme from Beeston, Nottinghamshire via the Somerset & Dorset at Templecombe. Down goods trains were limited to 12 wagons, while 15 could be conveyed in the up direction subject to a maximum load of 120 tons. However, for double-headed workings, the weight could increase to a maximum of 220 tons. In the declining years leading to the freight withdrawal on 3 February 1964, a single mixed train was more than ample to cope with the shrinking freight tonnage. Freight facilities were withdrawn from Combpyne on 5 December 1960.

Locomotives, no matter how good, do not last for ever, and by 1958 all three Radials were showing signs of old age. For all that, they could still show a thing or two to an interloper from another region, for on 11 and 12 November that year, an ex-GWR 0-4-2 tank, No 1462, arrived for trials. The first day was taken up with the engine running light and unfortunately the front coupled wheels did not take kindly to the severe curves. On the following day, it was booked to work two special morning trains, and

with the boiler proving incapable of supplying enough steam to prevent the brakes leaking on, the locomotive lost so much time with its first run of two coaches that the regular engine had to take over. And so the Adams locomotives were left to soldier on. No 30583 was given a general overhaul at Eastleigh Works early in 1959 and, on return, handled the bulk of the traffic, being the best of the trio. After reviewing a number of classes from other regions, an Ivatt Class 2MT 2-6-2T seemed to be the best option to be tested on the line, and after two of the curves were marginally eased and some lengths of track renewed in 1959-60, No 41297 worked a series of successful trials on 18 September 1960. However, because of the locomotives' weight at 63 tons, the class could only be accepted for emergency use. At this stage, the worst condition Adams Radial — No 30584 — was withdrawn in January 1961, having completed a total mileage of 2,102,781.

Summer Saturdays were still a problem with the expense of using the two Radials for the services, and with further trials using No 41308 proving that a

Class 2 could work the line unaided, permission was given by the powers that be for them to take over. Thus, by Spring 1961 the 2-6-2Ts were at work on the branch; the two Radials, Nos 30582 and 30583, were retained at Exmouth Junction as spare engines just in case of problems with the replacement motive power. No problems were encountered, however, and the 2-6-2 tanks worked the summer Saturday services unaided, bringing savings in operational costs. The two Radials were condemned in July 1961 and sent to the works at Eastleigh, where No 30582 was broken up in March 1962 with a mileage of 2,070,918. No 30583 was purchased by the Bluebell Railway and upon arrival at Sheffield Park was returned to LSWR livery, regaining its original number of 488.

Lyme Regis footplatemen worked the branch under Exmouth Junction Duty No 604, with the early crew booking on at 5.25am. The late crew would book on at 1.50pm and relieve their colleagues at 2.5pm, working the branch until the locomotive returned to shed at 9.23pm for disposal. The Ivatt 2-6-2Ts continued to work the branch for 2½ years until being replaced by two-car diesel multiple-units. Regular steam operation ceased on 4 November 1963 and single-car DMUs were in use in 1965. A shortage of diesel units caused a very brief reappearance of steam locomotives in the same year when an ex-GWR 0-4-2T appeared with a push-pull set on 15 February — this time no trouble was encountered. This was followed by 2-6-2T No 41291 on 20 February but this was only a short respite for steam operation and, following the closure of the Halwill-Torrington line in North Devon on 27 February that year, the single-car diesel unit used on this service was relocated for the Lyme workings, at busy times being supplemented with a driving trailer.

Footplate Memories

The replacement of the Adams Radial tanks by the Ivatt Class 2 2-6-2Ts was of a great benefit to the footplatemen; at last they had a more modern, free-steaming locomotive with a roomy, comfortable cab and the ability to drag five or six coaches unaided to Lyme Regis, although the Adams tanks still had the edge as far as grace and beauty were concerned. I will describe what it was like for the Lyme Regis crews to work with the Radials in tandem on a typical summer Saturday in the 1950s.

The sight and sound of two members of the class working together would bring photographers from far and wide to enjoy this unique spectacle, and today is no exception. We have five Bulleid coaches on our tail, packed full with passengers who have been left by the 10.45am from Waterloo which arrived at Axminster at 1.34pm. Remember, these are the days when people use the railways as a mass transit carrier. We are on the lead locomotive; the fire is built up and blazing away merrily — it always pays to fire the 'Billys' high up under the door with the back corners packed full and sloping down under the brick arch. The injectors are apt to be slow, so it is essential not to let the boiler glass go too far down, but at the same time not to overfill as the locomotives can prime. The heat in the cab is almost unbearable; the aroma of hot cylinder oil, coal smoke and steam hangs everywhere, the noise from our own and the other locomotive's vacuum exhaust echoes around the station. The twin needles in the vacuum gauge are quivering at the 21-inch mark and the steam pressure gauge is hovering just under the red mark. The whole cab is quivering ready for the off, the tall chimney of our sister locomotive breathing out smoke as the fireman prepares his fire. Up on our well-oiled footplate everything is ready in order to transport our passengers on the last few miles of the long journey from London to their holiday resort. Now it's time to depart. With a long blast on the whistle, which is answered by the locomotive behind us, my mate opens the regulator with the reverser in full forward gear. The first revolution of the driving wheels puts a tremendous blast on the fire, followed by another and another, with the blast from our exhaust repeating from the chimney behind us. The firedoor is now shut so as not to let cold air into the firebox.

The Adams locomotives have a softer exhaust sound than most locomotives, but they still put out a respectable blast as we pull up the 1 in 80 from the station and curve over the main line, whose shining ribbons of steel lie below us. Superb views of the valley of the River Axe are to be seen, but there is no time to gaze out from the cab as there is serious work to be done. We are now taking the strain of our five coaches as we hit harder into the ascent. We tackle the 1 in 66, which soon increases to 1 in 40, so it's shovelling time, and the sweat begins to roll down as the firebox is fed, a glowing white hot mass, blinding even on a hot summer's day. The injector is on to keep the boiler topped up; the whole cab shakes with power, while behind us, our comrades' chimney is throwing smoke and steam everywhere. The sound of our exhaust beats thunder around the countryside, sometimes in unison sometimes not, sparks and char stream skywards as we climb even further. Our coach flanges scream out in protest as we round the notorious Trill curves; a look rearward shows our five coaches being dragged slowly along. Here and there a head looks out of the windows, but soon disappears when the hot char from our chimneys drops earthwards. Higher and higher we struggle, the reverser is dropped further, the blast from the two exhausts is deafening; more coal is added to the firebox, and the firebox door slammed shut after each bout. Time for a quick breather by leaning over the side of the cab, not for long though, as no fireman can afford to relax at a time like this. The whole footplate is shaking with the massive forces in our cylinders,

there is steam everywhere, so wet you can almost wash your face in it — this is the steam locomotive at its best: noise, sight and smell. The exhaust from our sister locomotive darkens as the fireman attends his fire. The conditions for them are doubly bad, as most of our exhaust blast beats down on them. We blast under the overbridge near Hartgrove Farm, the smoke-stained portal bearing witness to years of struggling up the bank. Steam pressure drops as the injector is turned on again, and then rises again slowly. There will be no screaming safety valves on this trip!

We pass through the woods at Combpyne, well known for their rich carpets of bluebells in the spring, giving the branch the title of the 'Bluebell line', remembered even today by the local people. Although the incline eases slightly through Combpyne station, we are still climbing to the summit which is a little further east. We slog over the summit and the reverser is eased, as is the regulator. The injector is turned on again; there will be enough coal in the firebox for the time being. We can feel the weight of our train pushing us along as we clatter over Cannington Viaduct with its dip at the west end, the valley floor over 90ft below. The Radial noses around the sharp curve and, still heading downhill, we run past the village of Uplyme. Smoke whirls from the locomotive behind us as both locomotives are eased. It is time for a quick brush up of the footplate as our engine clanks along like a well-oiled sewing machine. The Lyme distant signal appears and my mate works the vacuum brake into the train with a loud hissing from the ejector; the twin needles drop and rise again as the ejector is pulled down and then released. We can feel the brakes binding on the train.

A loud blast on the whistle as we approach the station; we pass under the Lyme Regis-Axminster road bridge and more braking is applied as we come to a stand at the down home signal at 2.11pm with a groan from our own brake blocks and a squealing of brakes from the coaches. The blower is on to stop a backdraught into the cab; the interior is akin to a sauna with the heat from the firebox and escaping steam from the boiler — coal dust and sweat mingle on the skin. Here, we uncouple from the locomotive behind us, and enter the goods yard. The other Radial runs into the station with the train, and within a few minutes the platform is packed with passengers as they make their way to the station exit. Some of them will avail themselves of the waiting taxis, while most will walk down the hill into the town, complete with suitcases etc, en route to their various hotels and lodgings, in turn meeting folk walking in the opposite direction to catch the London train. Both locomotives are now watered and oiled in preparation for working the stock back to Axminster as the 3.5pm to Waterloo. The graceful Adams Radials were well known on other areas of the South Western, upon which they gave years of valiant service before they ever entered service on the line, but they will always be remembered by railway enthusiasts as *the* locomotives to be found working on the Axminster to Lyme Regis branch.

Left: Reflecting an age when classical design was the rule rather than the exception, Adams Radial '0415' class 4-4-2T No 30584, built in 1885, stands at Lyme Regis after arrival with the 8.43am from Axminster on 3 July 1956.
Hugh Ballantyne

SEATON BRANCH.
Summer Service, 18th July to 10th September only.—Weekdays.

To Seaton.

				—	1m. 49c.	2m. 41c.	4m. 16c.
Distances.				Seaton Junc.	Colyton.	Coly-ford.	Seaton.
				dep.	dep.	dep.	arr.
a.m.	Freight	...	...	7 30	{7 36 / 7 50}	7 54 / 7 58	8 4
„	Pass. ...	...	...	8 43	8 47½	8 50½	8 55
„	Pass. ...	...	...	9 36	9 40½	9 43½	9 48
„	Pass. ...	...	...	10 40	10 44½	10 47½	10 52
„	Pass. ...	NS	...	11 45	11 49½	11 52½	11 57
„	Pass. ...	SO	...	11 52	11 56½	11 59½	12 4
p.m.	Pass. ...	NS	...	12 28	12 32½	12 35½	12 40
„	Pass. ...	SO	...	1 0	1 4½	1 7½	1 12
„	Pass. ...	NS	...	1 5	1 9½	1 12½	1 17
„	Pass. ...	SO	...	2 5	2 9½	2 12½	2 17
„	Pass. ...	NFS	B	2 48	2 52½	2 55½	3 0
„	Pass. ...	FSO	B	3 15	3 19½	3 22½	3 27
„	Pass. ...	NFS	...	4 18	4 22½	4 25½	4 30
„	Pass. ...	FSO	...	4 28	4 32½	4 35½	4 40
„	Pass. ...	...	...	5 36	5 40½	5 43½	5 48
„	Pass. ...	NFS }	} A	6 41	6 45½	{6 48 / 6 51}	6 55½
„	Pass. ...	FO }	} B	6 51	6 55½	{6 58 / 7 1}	7 5½
„	Pass. ...	SO	B	7 3	7 7½	{7 10 / 7 13}	7 17½
„	Pass. ...	NS	...	7 40	7 44½	{7 47 / 7 50}	7 54½
„	Pass. ...	SO	...	7 52	7 56½	{7 59 / 8 2}	8 6½
„	Pass. ...	...	...	8 40	8 44½	{8 47 / 8 50}	8 54½
„	Pass. ...	FO	E	9 45	9 49½	{9 52 / 9 55}	9 59½
„	Pass. ...	SO	E	9 49	9 53½	{9 56 / 9 59}	10 3½

From Seaton.

				—	1m. 31c.	2m. 47c.	4m. 16c.
Distances.				Seaton.	Coly-ford.	Colyton.	Seaton Junc.
				dep.	dep.	dep.	arr.
a.m	Freight	...	...	6 0	{6 6 / 6 10}	6 14 / 6 34	6 40
„	Pass. ...	...	...	8 23	8 28	8 32	8 36
„	Pass. ...	...	...	9 10	9 15	9 19	9 23
„	Pass. ...	FO	...	10 0	10 5	10 9	10 13
„	Pass. ...	NF	...	10 14	10 19	10 23	10 27
„	Pass. ...	SO	...	11 15	11 20	11 24	11 28
p.m.	Pass. ...	NS	C	11 24	11 29	11 33	11 37
„	Pass. ...	NS	...	12 8	12 13	12 17	12 21
„	Pass. ...	SO	C	12 20	12 25	12 29	12 33
„	Pass. ...	NS	...	12 45	12 50	12 54	12 58
„	Pass. ...	SO	...	1 35	1 40	1 44	1 48
„	Pass. ...	NS	...	1 55	{2 0 / 2 2}	2 6	2 10
„	Pass. ...	FSO	C	2 35	2 40	2 44	2 48
„	Pass. ...	SO	...	3·50	3 55	3 39	4 3
„	Pass. ...	NS	...	3 55	4 0	4 4	4 8
„	Pass. ...	NS	...	4 56	5 1	{5 4 / 5 8}	5 12
„	Pass. ...	SO	...	5 0	5 5	{5 8 / 5 12}	5 16
„	Pass. ...	...	...	6 10	{6 15 / 6 18}	6 22	6 26
„	Pass. ...	NS	...	7 18	{7 23 / 7 26}	7 30	7 34
„	Pass. ...	SO	...	7 30	{7 35 / 7 38}	7 42	7 46
„	Pass. ...	NS	...	8 7	{8 12 / 8 15}	8 19	8 23
„	Pass. ...	SO	...	8 13	{8 18 / 8 21}	8 25	8 29
„	Pass. ...	FSO	D	9 0	{9 5 / 9 8}	9 12	9 16

A—Earlier times to be advertised.   B—Conveys through coach from Waterloo.   C—Conveys through coach for Waterloo.
D—To Axminster,   E—From Axminster.

SUNDAYS.—17th JULY to 25th SEPTEMBER only.

		Timing No.		Seaton J.	Colyton.	Colyf'rd.	Seaton.
				dep.	dep.	dep.	arr.
a.m.	Engine	...	A	9||35	...	...	9||50
„	Pass.	...	...	10 30	10 34½	10 37½	10 42
„	Pass.	...	...	11 22	11 26½	11 29½	11 34
p.m	Pass.	...	...	12 16	12 20½	12 23½	12 28
„	N.S.L.	238	AE	12 52	...	...	1 2
„	Engine	...	DE	12||52	...	...	1 ||2
„	Pass.	...	...	1 30	1 34½	1 37½	1 42
„	Pass.	...	E	2 57	3 1½	3 4½	3 9
„	Pass.	...	E	4 12	4 16½	4 19½	4 24
„	Pass.	...	E	5 24	5 28½	5 31½	5 36
„	Pass.	...	E	6 32	6 36½	6 39½	6 44
„	Pass.	...	E	7 32	7 36½	7 39½	7 44
„	Empty	...	A	7 45	...	{7 51 / 7 54}	7 59
„	Pass.	...	...	8 45	8 49½	{8 52 / 8 55}	9 0
„	Pass.	...	...	10 28	10 32½	{10 35 / 10 38}	10 43
„	Pass.	303	B	11 50	11 54½	{11 57 / 12 0}	12 5
mdt.	Pass.	303	C	12 8	12 12½	{12 15 / 12 18}	12 23

		Timing No.		Seaton.	Colyf'rd.	Colyton.	Seaton J.
				dep.	dep.	dep.	arr.
a.m.	Empty	...	A	9 0	...	...	9 15
„	Pass.	...	...	10 0	10 5	10 9	10 13
„	Pass.	...	...	11 0	11 5	11 9	11 13
„	Pass.	...	...	11 45	11 50	11 54	11 58
p.m.	Empty	...	...	1†15	...	...	1†25
„	Pass.	...	E	2 15	2 20	2 24	2 28
„	Pass.	...	E	3 40	3 45	3 52	3 56
„	Pass.	...	E	4 52	4 57	5 1	5 5
„	Pass.	...	E	6 10	6 15	6 19	6 23
„	N.S.L.	238	AE	7 0	...	...	7 10
„	Pass.	...	F	7 12	7 17	7 21	7 25
„	Engine	...	E	8 ||0	...	...	8||10
„	Pass.	...	...	8 12	{8 17 / 8 20}	8 24	8 28
„	Pass.	...	...	9 40	{9 47 / 9 50}	9 54	9 58
„	Empty	303	B	11†25	{11 30 / 11 33}	...	11†40
„	Empty	303	C	11†45	{11 50 / 11 53}	...	12 0

A—Runs 17th and 31st July, 14th and 28th Aug., 11th and 28th Sept.   B—Runs 24th July, 21st Aug. and 18th Sept. only   C—Runs 7th Aug. and 4th Sept. only.   D—Runs 24th July, 7th and 21st Aug., 4th and 18th Sept.   E—Exeter Engine.   F—Worked by Exeter engine on 24th July, 7th and 21st Aug., 4th and 18th Sept. only.

Winter Service, commencing 12th September.—Weekdays only.

				Seaton J.	Colyton.	Colyf'rd.	Seaton.
				dep.	dep.	dep.	arr.
a.m.	Freight	...	...	7 30	{7 36 / 7 50}	7 54 / 7 58	8 4
„	Pass. ...	...	...	8 43	8 47½	8 50½	8 55
„	Pass. ...	...	...	9 36	9 40½	9 43½	9 48
„	Pass. ...	...	...	10 40	10 44½	10 48½	10 52
„	Pass. ...	...	...	11 45	11 49½	11 52½	11 57
p.m.	Pass. ...	...	...	12 28	12 32½	12 35½	12 40
„	Pass. ...	...	...	1 5	1 9½	1 12½	1 17
„	Pass. ...	...	SO	2 17	2 21½	2 24½	2 29
„	Pass. ...	...	...	2 48	2 52½	2 55½	3 0
„	Pass. ...	...	...	4 18	4 22½	4 25½	4 30
„	Pass. ...	...	...	5 36	5 40½	5 43½	5 48
„	Pass. ...	...	...	6 41	6 45½	{6 48 / 6 51}	6 55½
„	Pass. ...	...	...	7 40	7 44½	{7 47 / 7 50}	7 54½
„	Pass. ...	...	...	8 40	8 44½	{8 47 / 8 50}	8 54½

				Seaton.	Colyf'rd.	Colyton.	Seaton J.
				dep.	dep.	dep.	arr.
a.m.	Freight	...	...	6 0	{6 6 / 6 10}	6 14 / 6 34	6 40
„	Pass. ...	...	...	8 23	8 28	8 32	8 36
„	Pass. ...	...	...	9 10	9 15	9 19	9 23
„	Pass. ...	...	...	10 12	10 17	10 21	10 25
„	Pass. ...	...	...	11 24	11 29	11 33	11 37
p.m.	Pass. ...	...	...	12 8	12 13	12 17	12 21
„	Pass. ...	...	...	12 45	12 50	12 54	12 58
„	Pass. ...	...	...	1 55	{2 0 / 2 2}	2 6	2 10
„	Pass. ...	SO	...	2 31	2 36	2 40	2 44
„	Pass. ...	...	...	3 55	4 0	4 4	4 8
„	Pass. ...	...	...	4 56	5 1	{5 4 / 5 8}	5 12
„	Pass. ...	...	...	6 10	{6 15 / 6 18}	6 22	6 26
„	Pass. ...	...	...	7 18	{7 23 / 7 26}	7 30	7 34
„	Pass. ...	...	...	8 7	{8 12 / 8 15}	8 19	8 23

Above: The Seaton branch workings for the summer of 1932 as taken from the SR Working Timetable for the year.

Seaton Junction-Seaton

A proposal to link the coastal village of Seaton with the main line station at Colyton for Seaton, using the name the Axminster, Seaton & Beer Junction Railway, was put forward in 1861. The railway was intended to run to Seaton only, in spite of the fact that the neighbouring fishing village of Beer was incorporated in the title, and the scheme involved constructing a bridge across the River Axe at Seaton. This proposal failed Parliamentary standing orders, but on 13 July 1863 an Act was passed to construct a line similar to the previous proposal, from the Colyton station of the South Western to Seaton, a length of 4¼ miles complete with a road bridge across the Axe, thus enabling the aims of a railway communication to the East Devon resort to be realised. A capital of £36,000 was authorised in shares of £10, plus a loan of £12,000. As always, the LSWR had a hand in the proceedings as it was to work the line for 45% of the receipts. The chairman, directors and officers of the company, whose title had now become the Seaton & Beer Railway, were elected at the first meeting which was held at the Pole Arms Inn, Seaton, on 5 December 1863. Sir Walter Calverley Trevelyan, Bart, of Nettlecombe in Somerset and Wallington in Northumberland, was appointed as chairman and George Evans Esq from Seaton as deputy chairman; the directors being William Dommett of Chard, John Babbage of Nettlecombe and John Latoysonave Scarborough Esq of Coly House, Colyton. The appointed officers were C. E. Rowcliffe, secretary;

Below: 'S15' 4-6-0 No 30827 stands on the up local line at Seaton Junction with the 3.20pm Exeter Central to Templecombe on 8 August 1960. *Terry Gough*

The curving branch platform at Seaton Junction on 24 August 1958. 'M7' No 30021 prepares to propel push and pull set No 381 to Seaton. The down local advance starter can be seen in the 'off' position in the background.
Terry Gough

Above: Adams Radial 4-4-2T No 30584 and Drummond 'M7' 0-4-4T No 30055 arrive at Seaton Junction with the 9am Seaton to Waterloo on 18 June 1949. *S. C. Nash*

W. R. Galbraith, engineer (who was also engineer to the LSWR); and Radcliffe & Davies, solicitors.

The contractor for the line, Howard Ashton Holden of London, signed the contract on 8 January 1864, and by 24 August of the same year it was reported that work had started. The course of the new line had been staked out from end to end, with a considerable saving in both land and works by a small deviation which had been endorsed by the Board of Trade. This had been requested by the S&BR who now required to run the line of railway along the east side of the Seaton Marsh bank by widening a portion of the bank, instead of the west side as originally proposed. However, by February 1865 little progress had been made, although notices for purchase of land had been served on all the owners concerned, including Sir John Pole through whose land the line would run for approximately two miles, the purchase price being £120 per acre, and with his agreement this had been completed on 2 January. The embankment running along the foreshore of the River Axe had been completed for approximately a mile near Seaton, but no progress had been made on other sections of the line. James Shipway, the agent to the contractor, being the recipient of complaints as to the tardiness of progress, reported 'the severe weather had affected the money market and had prevented more vigorous action on the civil engineering side'. Holden, for his part, had promised to proceed with the works rapidly, 'as soon as the weather has fairly set in'. On 23 April

the Seaton & Beer Railway instructed the solicitor and secretary to give notice to the contractor 'that unless on or before June 30th next he completes the works pursuant to Article 15 of his contract, the Directors will rescind and put an end to the contract and take such steps as counsel may advise.' This must have prompted Holden, for he then entered into partnership with a William Miller and, as a consequence, a supplemental contract was signed with Holden & Miller and the S&BR on 2 June 1865. Although the contractors were in possession of approximately four miles of line by August of the same year, again there was very little construction in progress. By now Galbraith, the engineer, had expressed complete dissatisfaction with the amount of progress, and believed that it would be impossible to complete the works by 1 January 1866 which was the deadline set by the supplementary contract and he advised that unless steps were taken at once, the summer traffic of 1866 would be lost.

Shipway found himself in the firing line again when he was informed that, unless active proceedings were taken within a fortnight, the contract would be terminated. No progress had been made three weeks later, with Holden & Miller found not to have the necessary finances, despite the fact that the S&BR Chairman, Sir Walter Trevelyan, had provided the contractor with a considerable sum, as well as taking half the shares. The contract was officially terminated on 27 September 1865. Another contractor, John

Above: With the regulator closed and smoke drifting lazily over set No 381, 'M7' No 30046 approaches Seaton Junction from Seaton on 24 August 1957. Note the LSWR lower quadrant signal arm. *A. E. West*

Below: The Seaton branch spare coach viewed at Seaton Junction on 24 August 1957. The vehicle is an ex-SECR 10-compartment third, No S1050S, in maroon livery. It is now preserved on the Bluebell Railway. *A. E. West*

Above: An evening milk train, headed by No 34033 *Chard,* departs from the up local line at Seaton Junction on 24 September 1964. The road ahead for the footplate crew is undemanding until Axminster, then begins the assault on Chard bank from Chard Junction until reaching the summit and descending into Crewkerne Tunnel. *A. E. West*

Below: 'M7' 0-4-4T No 30105, attached to the ex-LBSCR push and pull set, shunts milk tanks at Seaton Junction on 18 July 1948. *A. E. West*

Above: A six-wheeled Express Dairy milk tank, No B3184 in sea blue livery, at Seaton Junction on 7 March 1960. This wagon was built at Swindon to Lot No 1759 in 1950. *A. E. West*

Sampson, worked on the construction from 29 September, but despite his promise that within 10 days of acceptance he would place 20 earth wagons and 25 tons of temporary rail on the ground as plant, nothing happened and 11 days later, Radcliffe & Davies (the S&BR solicitors) became aware that the promise was hollow as Sampson was found lacking in wagons and temporary rail. Additionally, he had made no firm arrangements for the purchase of such materials, believing that he could obtain them on credit.

A meeting of the S&BR was held on 4 November 1865 when it was announced that the Railway Finance Company would be prepared to supply Holden & Miller with capital but, at the same time, if the directors were unable to await a fortnight, then Shipway would take over the contract himself. Unfortunately, he did not impress the directors with this proposal and was dropped from the proceedings. Holden's solicitors then wrote to the S&BR on 30 November informing them that as the railway company had threatened to let his client's contract, he would file a Bill applying for an injunction against the S&BR. To this the directors replied that they were willing to consider any proposals that Holden & Miller might make in a month. Meanwhile, pending any reletting of the contract, John Sampson carried on with construction of the line, but with at least three-quarters of the work still to be completed, a contract was signed with William Shrimpton on

17 February 1866 to complete the works at a sum of £26,500. This was to be made up of £8,000 cash, £5,000 in ordinary shares and £13,500 in 5% debenture shares. Just when everything seemed to be getting off the ground, William Shrimpton now declined to go ahead, claiming that Holden had threatened legal action against him if he proceeded with the contract. John Sampson, meanwhile, continued with the works, and a contract with him was successfully settled on 2 June 1866 for £12,000 preference stock and £22,890 in ordinary shares, with Sir Walter Trevelyan agreeing to purchase the shares and preference stock for £10,000 on the completion of the contract, and with Sampson agreeing to complete the contract by 1 March 1867. Tenders were now required for rails and sleepers, with Bird & Co tendering for the supply of 500 tons of rails at £6 10s 0d per ton, and William Wheaton of Exeter successfully tendering for 6,500 half-round sleepers at 2s 10½d each and 2,300 rectangular sleepers at 3s each. The Seaton & Beer Railway was itself by now in financial straits and it was realised that it had insufficient funds to further the line. On 18 April 1866 the Board of Trade authorised the issue of a certificate for raising additional capital of £12,000 in 5% preference shares and £4,000 on loan. The S&BR still had hopes of having the line ready by the summer of 1867, but at a half-yearly meeting in February of the same year it was stated that John Sampson had failed to make the progress expected.

Above: BR Standard Class 4MT 4-6-0 No 75001 shunts milk tanks on the branch side of Seaton Junction on 5 September 1964. *A. E. West*

Outside contractors also delayed matters, with Kerslake of Exeter refusing to supply tie rods, bolts and other materials for the uncompleted bridges until the S&BR guaranteed payment of £30. A locomotive was required in February 1867 for work on the line, but the one that had been promised could not be made available until April as it was out of action due to repairs, so a further six horses were used to bolster the resident horse power until a locomotive was hired from Isaac Watt Boulton's engineering works at Ashton-under-Lyne. This locomotive was urgently required to give assistance with the completion of the line as soon as the bridge across the Colyton turnpike was ready, and was running on the line by 2 August 1867, thus making the additional six horses redundant. Work was now progressing and attention was directed towards the station buildings, with a contract being signed with Birmingham & Co of Broad Clyst on 5 July 1867 for building the three stations at Seaton, Colyton and Colyford for a total of £6,700, plus a level crossing keeper's cottage at Colyton for a further £525. The same company was also responsible for constructing the wooden engine shed at Seaton for £200. The builders would have to get a move on as 1 September was the completion date for the booking offices and goods sheds, followed by the stationmaster's house one month later, as the opening of the line was still hoped to be on 1 October. By 2 August it was reported that the bulk of the earthworks had been finished, except for

the raising and widening of the embankments as required; if necessary the slopes could be trimmed and soiled after the opening. The culverts and masonry were virtually complete and ballasting had been completed on 2½ of the 4¼-mile-long track. Water, an important requisite for steam locomotives and humans at Seaton station, was originally to have been purchased from a Mr Hallett for the sum of £10 per year, with the railway laying a pipe under the River Axe for £50, but Hallett now limited the supply to 1,000 gallons per day, at the same time refusing any water for station use at Seaton. The contract with him was abandoned, and for £10 per year the railway obtained a satisfactory supply from the estate of Sir Walter Trevelyan, the chairman of the Seaton & Beer Railway. It was also reported that the Electric Telegraph Company had installed its apparatus along the length of the branch for £150.

The Board of Trade was notified that the railway would be ready for inspection by 20 December 1867, and Colonel Yolland arrived on 27 December. The line, 4 miles and 30½ chains long, was constructed of 75lb flat-bottomed rail laid on fir or larch creosoted sleepers, the ballast consisting of flint gravel and shingle from the shore at Seaton. 10 underbridges and three viaducts were provided, with stations at Seaton, Colyford and Colyton. The Inspector did not sanction opening of the line on this occasion for various reasons. He was not happy with the branch arrangements at Colyton Junction where branch

Above left: The down advance starting signals with LSWR lower quadrant arms for the local and through lines respectively are viewed at Seaton Junction on 12 March 1952. A sextet of six-wheeled milk tanks stand on the up sidings in the background. *A. E. West*

Above right: SR upper quadrant signals guard the exit from the branch platform at Seaton Junction on 19 April 1963. Graded in height according to its importance — the signal to the right is the starting signal to the down main line from the branch. The signal on the left provides access to the down sidings. The signalbox dated from 1928 and contained a 55-lever frame. *A. E. West*

trains had to be shunted into and out of the down platform at the station with passengers aboard for a distance of approximately 200 yards; he found this idea objectionable and believed it would be more desirable to construct a platform alongside the branch line adjacent to Colyton Junction station. In addition, the signalling arrangements, including the locking apparatus, at Colyton Junction were not complete; the platform was too short at Colyford station (it needed to be extended by another 50ft), and the check rails at the crossings at the junction station and Seaton were not placed at the proper distance from the rails. In fact, the Inspector's report

contained eight paragraphs of items to be attended to, including one that stated, 'clocks that can be seen from the platforms are required at all stations'. However, Colonel Yolland returned for a reinspection on 19 February 1868, and reported to the Board of Trade that all his requirements had been attended to except for the first and sixth, the former being the branch arrangements at Colyton Junction, and the latter being the need for additional ballast for binding purposes where the ballast was wholly of shingle. In fact, Sir Walter Trevelyan, during an interview with the President of the Board of Trade, was informed that the line could be opened subject to the undertaking that a platform at Colyton Junction would be built within six months.

The directors of the Seaton & Beer were shocked and upset after receiving a letter dated 27 February 1868 from Captain Charles Mangles, Chairman of the LSWR, requiring the S&BR to set aside money so that the timber-built engine shed at Seaton, the platform walls at Colyton and two timber bridges could all be replaced, and also that transverse wrought-iron girders could be inserted in other bridges. All this and the line had not even opened yet! The directors of the S&BR went to the President of the Institution of Civil Engineers, as they were justified in doing, and sought arbitration but to satisfy and keep the South Western appeased during the following month, the S&BR gave an undertaking that it would replace the structures with durable materials. In a judgement dated 15 March 1869, it was stated that the S&BR was properly constructed and that the company should rebuild with brick, stone or iron the timber-built bridge situated north of Colyford, but that this bridge was anticipated to last at least 14 years from the opening of the line with the replacement costing £200. Satisfaction was expressed, however, with the timber bridge to the south. A covered way was also required to be constructed between the platform waiting shed and the booking office at Seaton. The line was at last opened on 16 March 1868 and the main line station renamed Colyton Junction from the same date. *The Times,* published the following day, reported: 'This branch from the Colyton Junction on the South Western Railway to Seaton was opened for traffic yesterday. It is about 5 miles in length and cost about £35,000. The line runs through Colyton and Colyford to Seaton, close to the Axmouth Harbour. Sir Walter Trevelyan, Lord of the Manor of Seaton, is the principal shareholder and the line has been leased to the South Western Railway Company.'

Although there were no joyous public celebrations to mark this event, as happened at other locations, nevertheless many inhabitants took the chance to travel on the railway that day. Services from the beginning totalled four down passenger trains plus a mixed train, and three passenger and two mixed services in the opposite direction but with no trains run on Sundays. The line was originally worked on the train staff system as 'one engine in steam'. This method of working changed to the Tyers tablet system from Sunday, 5 March 1899 with an intermediate signalbox at Colyton which, however, was seldom used. Beattie 2-2-2 well tanks of the 'Tartar' class, No 12 *Jupiter* and No 33 *Phoenix*, worked the trains from the opening, the locomotives being stabled at Seaton. The line ran nearly due north from Seaton, separated from the road to Axminster by the River Axe and marshland, quite inaccessible except to birds and wildlife. After three quarters of a mile, the road turned eastward away from the estuary towards the village of Axmouth, leaving the line to

continue on the west side of the valley of the River Axe. This geographical feature then divided, leaving the railway to travel through the valley of a tributary — the Coly — before reaching Colyford station, approximately 1 mile 31 chains from Seaton. The line from here, guarded by level crossing gates, crossed the Exeter-Lyme Regis road and climbed at 1 in 168 for a quarter of a mile and then at 1 in 76 to Colyton (2 miles 47 chains). Leaving Colyton on a short stretch of 1 in 300, there was a further ¼ mile at 1 in 76 which carried the line over the Colyton-Axminster road with the gradient almost continuous at 1 in 100 until reaching the main line junction (4 miles 16 chains).

In the original Act of 1863, the Seaton & Beer Railway had powers to construct a toll bridge across the mouth of the Axe to accommodate traffic from Axmouth and Rousdon to Seaton station. Prior to the bridge being built, the inhabitants of Axmouth had to use the station at Colyford. A contract for building the bridge was signed with William Jackson of Westminster on 15 December 1875, with Philip Brannon as the engineer. The bridge, with a central span of 50ft plus two side arches of 30ft, spanned the main stream of the river and the tideway and is believed to be one of the first bridges in the UK to be constructed in concrete, with the adjoining toll house being the oldest concrete house in England. Opened to public traffic on 24 April 1877, the tolls were 4d for each horse and cart, 1d a leg for animals in harness, and ½d a leg for animals not in harness, the staff of Seaton station being allowed to cross without charge. The LSWR and the GWR were approached by the S&BR in February 1879 to ascertain what terms they would offer for the purchase of the line. A reply from Archibald Scott, the Traffic Manager of the LSWR, stated that the company would take out a 1,000-year lease for the sum of £1,000 per annum, increasing by annual instalments of £100 to £1,500, and in the following year rising by £50 to a sum of £1,550 per annum in perpetuity, thus allowing an income of 4% on debentures and the arrears of the same, 3% on preference shares and 1% on ordinary shares. The South Western would also have the option of purchasing the S&BR by issuing to debenture holders and shareholders a sufficient quantity of its own 4% preference stock to meet the respective dividends, with the result that on the receipt of such stock the Seaton & Beer company would cease to exist. The S&BR agreed to the tentative terms of the South Western, although the larger company was not interested in purchasing the toll bridge across the Axe, and terms were arranged by the S&BR for transfer of the bridge to Sir Alfred Trevelyan of Nettlecombe (the successor to Sir Walter Trevelyan), the deed of transfer and conveyance being completed on 27 August 1881, with the bridge eventually being made toll free in 1907. The LSWR acquired powers to purchase the Seaton & Beer company on 26 August 1880, with acquisition taking place on 1 January 1888.

Above: Set No 603 arrives alongside the branch platform at Seaton Junction as empty stock from Seaton — hauled by Drummond 'M7' 0-4-4T No 30048 on 8 August 1960. The spare Seaton branch coach stands on the siding to the right. *Terry Gough*

Seaton Junction

Lying just three miles west of Axminster in the valley of the River Axe, 'Colyton for Seaton' opened with the main line to Exeter Queen Street on 18 July 1860 as a single-line station. As with Yeovil Junction, Chard Road and Ottery Road, it was located some way from the larger centres of population, and although located near the village of Shute, was instead named after the village of Colyton, situated some two miles from the station. Shute is well known as the location of Shute Barton, now owned by the National Trust, one of the most important non-castellated houses of the Middle Ages. The building of Shute Barton began in 1380, and it later became the property of the Grey family, Dukes of Suffolk. They lost it, and their heads, in 1554 when the unfortunate family made their bid to place Lady Jane Grey upon the throne. Prior to the opening of the branch, the normal route for visitors travelling to and from Colyton and Seaton was via Axminster or Colyton for Seaton (as it then was known), and thence by horse conveyance. The Shute Arms Hotel, just outside the station, was constructed in 1898 as a result of a brewery executive having to travel to

Chard and Yeovil, and foreseeing the need for a stabling point to house the horses for such passengers. With the opening of the line to Seaton on 16 March 1868, the junction station became Colyton Junction with the new station on the branch being named Colyton Town. The junction was renamed from July 1869 as Seaton Junction in order to avoid conflict with the new station on the branch, which itself had 'Town' dropped from its title in September 1890. The main station buildings were situated on the up platform and consisted of the usual Tite style, as at other locations on the route, constructed in dark coloured brick with ashlar dressings, stone facings to corners and windows, and prominent gables. The chimneys were unusually wide, however, when compared with other similar structures along the Salisbury to Exeter line. The building contained the stationmaster's house and station offices and there was a canopy shelter protected waiting passengers from inclement weather. A goods shed and cattle pens etc were to be found on the up side of the layout. A footbridge linked the main line platforms; the down platform was unusually equipped with a hip-roofed shelter-cum-waiting room, and a separate canopy shelter, which was positioned further west along the platform, was intended for Seaton branch passengers.

Above: 'M7' No 30048 propels the 2.55pm to Seaton away from Seaton Junction on 8 August 1960.
Terry Gough

Below: No 30048 again, with a good stack of coal in the bunker, propels set No 603 and a through coach along the down main at Seaton Junction on 8 August 1960. Thence it reversed to the branch platform and worked the 4.47pm to Seaton. *Terry Gough*

The original signalbox was situated off the Exeter end of the platform. Seaton Junction had no platform for the branch itself, trains having to depart from a bay on the down side, running in to a siding and then reversing to travel down to Seaton. The opposite procedure had to be performed for arrivals from the branch, and this inconvenient method of operation was to exist until the rebuilding of the station in 1927-8. Even after the branch was opened, it was common for main line trains to omit calling at Seaton Junction, and for the branch train to travel to Axminster and back to make the connection — Axminster appeared in the branch timetable for many years.

From the Appendix to the Book of Rules & Regulations and to the Working Timetables. 1 January 1911.

Special Instructions for Seaton Line. 'During certain portions of the year, this line is worked by a small type of engine and the loads of trains must then be arranged as follows:
Maximum Loads Seaton Junction to Seaton

Passenger trains:	Equal to 40 wheels, including brake vehicles
Mixed trains:	Equal to 52 wheels, the passenger stock on a train not to exceed 24 wheels
Goods trains:	13 wagons and 1 van
Passenger trains:	Equal to 40 wheels including brake vehicles
Mixed trains:	Equal to 40 wheels, the passenger stock on a train not to exceed 24 wheels
Goods trains:	11 wagons and 1 van to Colyton
	10 wagons and 1 van Colyton to Seaton Junction.

Left: LSWR lower quadrant signal arms give a clear road ahead for 'H15' 4-6-0 No 30335 (a Urie rebuild of Drummond 'E14' class No 335 and subsequently modified by Maunsell) departing from the up local line at Seaton Junction on 24 August 1957. The spectacular layout with the through passing tracks in the centre is seen to good effect in this view. *A. E. West*

Below: The branch set, plus a box van for loading from the egg depot which had direct access to the platform from the building under the short canopy seen here just behind the locomotive, has left alongside the up platform at Seaton Junction on 2 September 1959. 'M7' No 30045 pulls forward to make a shunting movement. The milk tank wagon on the left is on the private siding to the Express Dairy milk and egg depot housed in the former goods shed. *H. B. Priestley — Author's collection*

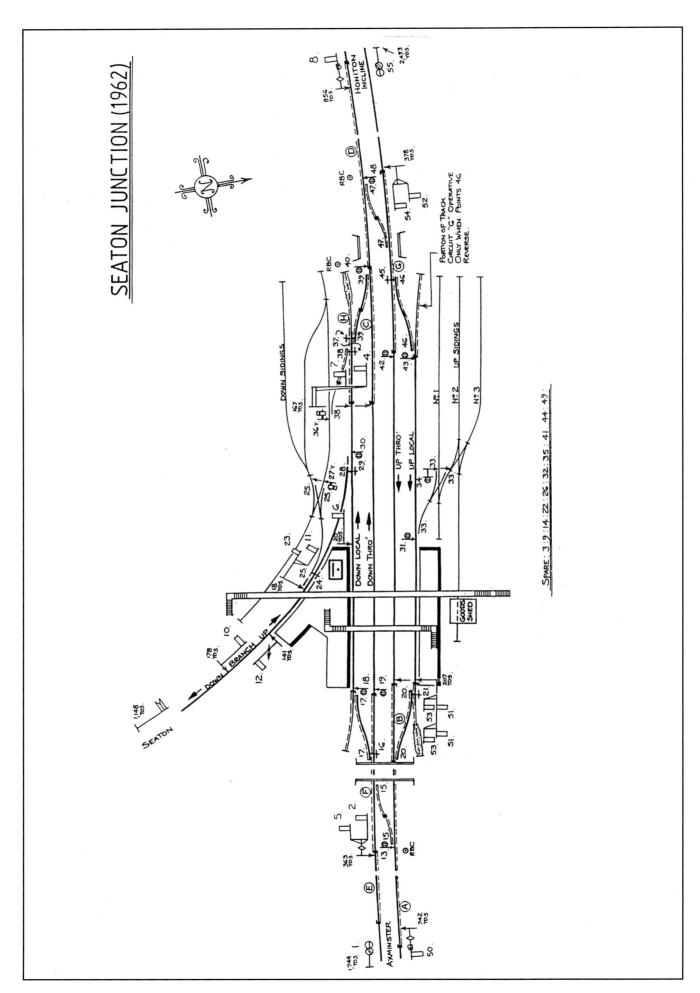

SEATON JUNCTION (1962)

Above: Collett 0-4-2T No 1450 stands at the Seaton Junction branch platform on 20 February 1965. This locomotive, originally numbered No 4850, was built by the GWR in July 1935 and renumbered in November 1946. *S. C. Nash*

The Southern Railway reported at the 1925 Annual General Meeting its decision to rebuild the station and provide passing loops at a cost of £46,000. The works were considerable, and although the up platform (albeit lengthened and with an improved extended canopy), station buildings and goods shed were retained, it was the down side of the layout that was to witness the major alterations. In order to enable the two additional tracks to be provided, the down side platform and cast-iron footbridge were removed, with the new down platform now extending around the curve of the branch. With the new branch platform coming into use on 13 February 1927, branch trains no longer had to reverse in and out of the station. The new down through line was brought into use on Sunday, 1 April 1928, and the up through line on Wednesday, 4 July. The new loops provided the only passing place on the 49-mile stretch of line between Yeovil and Exeter. A new 55-lever signalbox replacing the original opened with the new layout on 3 April 1928, situated on the Exeter end of the new down platform. Two new footbridges, constructed in the Exmouth Junction concrete style, were also provided, one of which was a public right of way linking the village of Shute to Lexhayne Farm over the station platforms and tracks. An attractive feature of the layout was the tall co-acting up starting signals, which remained lower quadrant until

demolished in 1965. The signals themselves were unusual in the fact that the usual practice in co-acting signals was for the lower arms to follow the exact priority as the upper arms; however at Seaton Junction the upper arm for the up through line was higher than that for the loop, while both lower arms were of the same height. The former goods shed at the rear of the up platform was converted into a cooling shed for milk in April 1934 with rails extending into the premises for 30ft. Water, a prerequisite for the cooling operation, was at first obtained by sinking a borehole to a depth of 280ft, although water was eventually supplied from a depth of 34ft. With a 50hp oil engine pumping 90 gallons an hour it was not satisfactory, although at the time it was better than nothing. The first milk went through the plant during the second week of September and was despatched in a GWR six-wheeled tank carrying the name West Park Dairy Co Ltd, but only two-thirds full instead of its 3,000-gallon capacity. The Express Dairy became the new owner on 1 October 1934. New contracts enlarged the production to 4,000 gallons, the cargo being sent out in one tank wagon and the remainder in churns. The water supply was improved when the Southern Railway, at the expense of the milk company, constructed a 3in pipeline from Honiton Tunnel to the station, a distance of five miles, the water also supplying the station as well as the dairy. The Express Dairy also paid for the power lines to be established overland to the junction.

Before the through coach to Waterloo (10am ex-Seaton) was withdrawn in the autumn of 1962, it was attached to the 8.25am from Seaton and, upon arrival, the branch stock was left at the junction, the through coach returning to Seaton thereby enabling it to be steam heated for the passengers awaiting the service to Waterloo. Goods work on the branch was originally handled by mixed trains; however, from 1919 a goods train plus two mixed services and conditional goods were provided for in each direction. Wagons for the branch in 1963 were detached from down main line trains at Axminster rather than Seaton Junction; the Lyme Regis locomotive collected its own share, and a locomotive, usually a 2-6-2T, was sent from Exeter for the Seaton goods, leaving at 7.15am and returning to Seaton Junction light engine. Trains to Seaton were down and up in the reverse direction. With the replacement of milk churns by rail-borne tankers from 1931, Seaton Junction, as with Chard Junction and Semley, quickly became a major distribution centre for such traffic until the mid-1960s.

Seaton Junction was closed to passenger traffic on the same day as the branch, on 7 March 1966. This was a black weekend for railways in the West Country, with not only many of the intermediate stations closing between Salisbury and Exeter, but the complete closure of the Somerset & Dorset as well. General freight facilities were withdrawn on 18 April the same year, but coal traffic continued until 8 May 1967. The signalbox also closed in 1967, on 11 June, the main line being singled from the same date with all trains using the former down through road. Milk traffic continued for a short time after, and in October 1972 the single line was slewed over to the site of the former up through line and a siding occupied the site of the former up loop until the 1980s.

Today, as the Class 159 Turbo units run through the remains of this once busy junction, only memories exist of crack trains racing through on the centre tracks, the little Drummond 'M7s' awaiting the next journey down the branch with the Westinghouse pump on the side of the smokebox panting away as it built up air pressure for the driving trailer, or of the hundreds of passengers tramping across the footbridge with the chance of a cup of tea before the arrival of their train to London. The former down and branch platforms mostly survive, although the latter has been cut short, and it is still possible to trace the outlines of the station building and signalbox on the platform. The two distinctive footbridges remain and the buildings on the up platform are complete, including the canopy structure and the small store on the Honiton side of the footpath bridge.

Minute 6710 (v) Traffic Officers Conference 7 October 1929. Seaton Junction 26 August 1929. '

At 5.55pm, when shunting operations were being carried out at Seaton Junction with the 5.20pm milk train from Sidmouth Junction to Waterloo, a portion of the train was set back with some force against the stop blocks at the dead end of the up local line at the Honiton end of the station, resulting in brake van No 791 being derailed and damage caused to two other vans... Signalman Reddaway expresses his regret for his temporary lapse in omitting to set the points, he has been reprimanded and… Porter Signalman Love has been reprimanded.'

Minute 6830(j) Traffic Officers Conference, 16 December 1929. Seaton Junction 27 November 1929.

'At 7.25pm, when shunting operations were being carried out at Seaton Junction, tender engine No 829 was erroneously admitted to the short dead end siding at the western end of the down local line, where it mounted and demolished the stop blocks and dropped down the embankment, blocking the public by-road which passes underneath the railway. The tender was also derailed and the down local line obstructed … Signalman Bunce has been suspended from duty for three days and Relief Porter Hooper has been similarly dealt with, a shunt signal is provided in the down main line at the trailing end of No 39 connection, applicable to movements on the down local or down through line, and the question of the driver's responsibility is being followed up.'

Right: The station staff gather at Colyton for the photographer. The cross on the signal arm denotes that it is out of use. This was often the case at Colyton as the signalbox was only switched in for the peak summer timetables. Signalling came to the station in 1899 and was removed in 1922 when the box was reduced to a ground frame released by the single-line tablet. The goods shed stands to the left, with the station platform on the right. *Lens of Sutton*

Colyton

Colyton was once host to a Saxon parliament held in AD827 by Egbert, King of Wessex, and is mentioned in the Domesday records of 1083 as forming part of the West Saxon Royal Demesne. By the 16th century it was a prosperous wool town, owned mostly by Henry Courtenay, Marquis of Exeter who had the misfortune to fall out with none other than King Henry Vlll, not only losing his lands in the process, but his head as well! The town, with its mills, tanneries and stock fairs, was at one time a far more important centre for trade and industry than Seaton.

The single platform station was situated on an embankment just east of the River Coly and was located approximately half a mile from the village. The suffix 'Town' was dropped from the station title in September 1890. The extensive main station building, which contained the stationmaster's accommodation as well as the station offices, was constructed in red brick. A distinctive feature of the building was the provision of double sliding doors from the platform to the booking office. A small detached brick-built goods shed and store was situated at the up end of the platform, a water tank was provided at the Seaton end of the platform, and two sidings and a goods shed lay on the west side of the line, with the far siding standing adjacent to a stone store owned by the well-known firm of Messrs Bradford & Sons. The signalbox, dating from 5 March 1899, enabled the block section to be split into two sections. Signalling consisted of two fixed distants, a down inner home and a down station starter, plus an up outer home and an up starter signal. Two ground signals were also provided for entry and exit to the goods yard. Even after the box was opened, it was switched out for periods of time, its signals fitted with large white crosses denoting them to be out of use. The box was brought into use at peak times in the holiday period, when the Seaton Junction to Seaton tablets were locked away at Seaton Junction and a fresh set, retrieved from storage by a key held by the District Superintendent at Exeter, covered the two separate sections Seaton Junction-Colyton and Colyton-Seaton.

1911 Appendix to Working Time Table Special Instructions for Intermediate Sidings — Seaton Line

Colyton station. 'The goods sidings at this station is [sic] situated on the up line side and the points, which are facing for down trains, are worked from the signalbox and controlled by the tablet for the Seaton Junction and Seaton Section in accordance with the regulation for controlling sidings by means of the electric train tablet.'

The signalbox was abolished as a block post and reduced to a ground frame on 4 April 1922 with all signals removed, the box finally closing on 11 November 1958 and being replaced by a two-lever ground frame operated by a key on the single-line token to gain entry to the sidings. Stationmasters over the years included J. Wyeth, H. E. Millichap and H. W. Hodges until 1928 when the station came under the control of the Seaton stationmaster. The station became unstaffed on 3 February 1964 with the goods yard closing on the same day, trackwork in the yard having been lifted by 19 May of the same year.

Below: The 1.37pm from Seaton arrives at Colyton formed of 'M7' No 30021 and set No 381 on 6 July 1959. *H. C. Casserley*

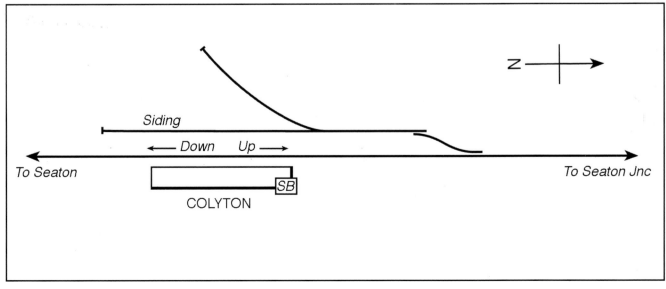

Above: Colyton, looking towards Seaton, on 7 October 1954. Standing adjacent to the signalbox is a small building with the sliding doors which was used as a goods store. The station buildings fit neatly into the centre of the single platform with the water tower standing at the end with the goods yard and shed, with wagons for the local traders shunted by up trains from Seaton on the right. Today the station is the terminus of the Seaton Tramway. *J. J. Davies*

Below: Colyton, looking towards Seaton Junction, again on 7 October 1954. The fencing panels were a product of the SR's concrete works at Exmouth Junction. *J. J. Davies*

Above: A 12-ton LNER van, No 081141, branded 'Commercial Colyton', stands in the goods yard on 25 June 1962. The roof of this vehicle must have been leaking — hence the tarpaulin sheet. *A. E. West*

Below: Colyton, looking towards Seaton on 7 July 1959. Box vans stand on the goods shed road and the station buildings bask in the summer sunshine while a couple of passengers wait for a train. *H. C. Casserley*

Above: A rear view of the driving trailer of set No 381 hauled by 'M7' No 30021 leaving Colyton with the 1.37pm from Seaton on 7 July 1959. *H. C. Casserley*

Colyford

Colyford has been a royal borough since the reign of King John. A mayor still holds office and at one time it even had a Member of Parliament. The annual Goose Fair, for which the charter dates back to 1340, is a popular event, taking place in late September each year. The small station was the most modest on the branch and was no more than a single platform halt located on the west side of the line, serving the village of the same name on the River Axe and not being provided with sidings or goods facilities. Accommodation was altered over the years, but originally comprised a timber-built single-storey booking office and waiting room; unfortunately this small but attractive building was removed and replaced by a precast concrete hut — austerity indeed! The only other additional 'building' was a cast-iron gents' toilet which is still extant today. A level crossing, situated north of the station, protected trains crossing the busy Exeter-Lyme Regis road.

The staff comprised a single porter, and when no member of the staff was on duty it was the responsibility of the guard to collect tickets and open and close the level crossing gates. When a down train pulled up at the stop signal, he would be responsible for lowering the home signal to admit the train to the platform and, when the train had cleared the crossing, had to place the signal back to danger, all of this being accomplished by using the ground frame. If the train was the last of the day to use the station, then the guard had to extinguish the platform lamps. The signalling at Colyford originally consisted of two fixed distant signals with a pair of LSWR lower quadrant signals guarding the crossing itself, which were operated from a ground level cabin housing a six-lever ground frame. All the signalling was removed in February 1949 and the distant signals replaced by marker lights, with levers Nos 1, 2, 3 and 4 being made spare. Further alterations were made in November 1957 when a new two-lever ground frame appeared, with No 1 lever operating the gate lock and No 2 locking the wicket gates. A telephone was provided at Seaton in its latter days especially for the

Above: Bulleid Pacific 4-6-2 No 34096 *Trevone* approaches Colyford with the 10am Taunton to Seaton excursion on 6 June 1960. *S. C. Nash*

use of the guard who, before a train's departure from the terminus, would telephone the crossing keeper at Colyford that his train was about to leave. If unsuccessful in contacting the crossing keeper, he would then instruct the driver to approach the crossing with caution and expect to find the gates closed against the train. Light engines were prohibited from travelling on the Seaton branch when the porter at Colyford was not in attendance, unless they were accompanied by a member of the traffic staff, who would then attend to the opening and closing of the crossing gates.

Minute 254 (a) Traffic Officers Conference, Tuesday, 21 July 1913 Colyford Station Level Crossing. '

At about 1pm, the 12.44pm excursion train from Seaton Junction to Seaton passed the signal at danger and ran through the level crossing gates which were closed across the line at the time, and demolished same... Driver Wilson, who was responsible for the mishap, has been suspended for one week.'

Colyford had its own stationmaster until 1927 when it came under the control of the stationmaster at Seaton. Stationmasters up until 1927 included G. Godber, W. Rowden and Mr Littley. Unfortunately, in today's society reports of vandalism and general thievery are commonplace but scoundrels were also at work in the 1920s — hence the following incident that happened on 27 June 1929.

Minute 6673 Traffic Officers Conference 22 July 1929 'Burglaries on the Company's Premises. Colyford Booking Office, 27 June.

'The outer door lock was broken and the inside door forced, the screws of the lock being wrenched off. An amount of £2 1s 4d was missing and a suitcase, the property of a Major Woodcock of Colyford, was stolen.'

Passenger tickets issued at Colyford in 1928, 1932 and 1936 totalled 4,698, 2,261 and 2,385 respectively, with six season tickets being issued in the years of 1932 and 1936.

Above: The driving trailer of push and pull set No 373 can be seen beyond the Colyford level crossing in this 1920s view. The driver is in the cab and is presumably waiting for the guard to open the gates in the absence of a member of the station staff. The ground frame for the signalling and unlocking the gates can be seen to the right by the wicket gate. Part of the timber-built booking hall, complete with paraffin lamp, can be seen to the left. *Lens of Sutton*

Below: Colyford station, looking north on 7 October 1954. Newly laid trackwork extends to the level crossing gates, with the old rails lying discarded. The cast-iron gents' can be seen on the platform. *J. J. Davies*

Above: Colyford as viewed towards Seaton, with a branch train in the background. *Lens of Sutton*

Below: Colyford, looking towards Colyton, on 7 July 1959. A member of the station staff has just placed the level crossing gates across the road in the expectation of an approaching train. The booking hall has been removed, leaving the gap in the concrete fencing, left. *R. M. Casserley*

Above: Seaton station appears more in keeping with a farmhouse than a railway terminus. An early view of the original station building which was constructed by Birmingham & Co of Broad Clyst and opened on 16 March 1868.
Lens of Sutton

Seaton

The station at Seaton was sited on the west bank and close to the mouth of the River Axe, opposite (although separated by a ridge of high ground from) the sea front and its shingle beach. It was the proximity of the station to the beach that made it one of the most popular resorts for day trippers, compared with Lyme Regis or Sidmouth. However, passengers walking from the station had to face the daunting aspect of passing the goods yard entrance and gasworks in order to reach the facilities of the town centre and, as the town grew, so the inhabitants and trippers had a further walk to the amenities offered in Seaton Square and other areas of the town. The station at first consisted of a very short wooden platform with two faces, one used for the branch services and the other for stabling coaching stock. Accommodation for the stationmaster took up a considerable part of the station; a goods shed and crane were provided in the small yard, and a cattle pen was situated at the terminal end of the station. Clarke, the LSWR secretary, wrote to the S&BR on 28 July 1869 stating that, in accordance with the agreement of 31 December 1867, the platform at Seaton should be extended northward for a further 180ft to enable it to cope with excursion traffic. The lengthening was carried out in the spring of the following year at a cost of £60 3s 0d. Seaton was for many years an 'open' station, with down trains having to stop at a ticket platform situated outside the station to allow tickets to be collected, this arrangement lasting until the Grouping. In the absence of a run-round loop, trains had to reverse out of the station to allow locomotives to change ends. The LSWR installed a 5-ton crane in the yard in 1872.

The only major source of goods traffic on the branch was the Seaton gasworks, situated west of the station, and at the 1909 Annual General Meeting of the Seaton Gas & Coke Co, the chairman reported that plans for a private siding had been agreed with the LSWR, but that an alternative proposal from the South Western was being considered. The siding was never provided. Statistics for coal and coke received at Seaton in 1928, 1932 and 1936 were 5,007, 6,073 and 5,832 tons respectively. A large travelling overhead crane was installed at Seaton to deal with the Beer stone traffic which was brought to the station by road; unfortunately the stone traffic failed to live up to its early promise, and the trade grew less and less until, eventually, the overhead crane was removed when the station was rebuilt.

Weekday Service July 1868 Public Timetable

Down Trains	am		pm				
Colyton Junction	8.50	11.20	12.40	3.18	5.30	7.55	8.55
Colyton Town	8.57	11.30	12.48	3.26	5.38	8.02	9.02
Colyford	9.01	11.35	12.52	3.30	5.42	8.06	9.06
Seaton	9.08	11.45	12.58	3.36	5.48	8.13	9.13
Up Trains July 1886	am		pm				
Seaton	8.00	10.35	12.00	1.30	4.20	7.25	8.25
Colyford	8.05	10.42	12.06	1.37	4.27	7.32	8.32
Colyton Town	8.09	10.46	12.16	1.41	4.31	7.36	8.36
Colyton Junction	8.15	10.54	12.24	1.49	4.39	7.44	8.44

With the opening of the line, passengers could depart from Seaton on the 8am service and catch the 8am from Exeter Central at Colyton Junction at 8.40am with a booked arrival in Waterloo at 2.13pm. Down services from the capital were swifter; using the 7am from Waterloo, upon which first, second and third class coaches were provided, travellers could reach Seaton at 12.58pm, or 5.48pm if using the 11.40am departure from London. First and second class passengers could arrive at Seaton at 3.36pm and 9.13pm by using the 10.50am and 3.50pm departures from Waterloo respectively. The single fares to Seaton from Waterloo in the opening year were 30s 7d (first), 22s 3d (second) and 12s 7½d (third), a three-day return costing 49s 6d (first) and 36s 6d (second). There was no Sunday service on the branch. Seven passenger trains ran each way in August 1874, the journey averaging approximately 17 minutes, with one making a return trip to Axminster and back, plus a mixed train making a return journey along the branch.

Below: The original station at Seaton with the building and stationmaster's house in the background, showing the basic platform shelter and the lack of run-round facilities for locomotives. Trains had to be reversed out of the station in order for locomotives to be released. *Lens of Sutton*

Summer excursion traffic on the branch seems to have begun almost as soon as the line was opened, with a local paper reporting an incident which happened on a Tuesday evening in September 1871, the main line Seaton-bound excursion train running into the level crossing gates situated between Crewkerne station and the tunnel. Some 5,000 passengers arrived by rail on Whit Monday 1909, including 526 from London and 835 from Chard, this influx straining the limited siding accommodation to the limit! The fare for a day excursion from Waterloo, departing at 6.30am and arriving at Seaton at 10.55am, was 7s 6d return (third class). The summer working timetable for 1909 shows a GWR excursion from Taunton via Chard Junction running every Wednesday from 7 July until 15 September inclusive,

with the timetable also proclaiming nine passenger trains, two mixed and one goods in each direction, plus an early morning conditional goods. The LSWR's Summer 1914 Excursion Train Notice shows that the GWR engine worked through to Seaton, with an LSWR pilot driver from Chard Junction. The official guide to Seaton and District 1919 observed: 'The Great Western Railway runs excursions into the town via Chard, and the influx of holidaymakers on Regatta Day and other notable occasions is very great but the beach covers such an extensive area, that there is no risk of anything like crowding and even on a busy day, resident visitors need be under no apprehension as to the tripper element.' The fare for a Bank Holiday Monday excursion on 6 August 1962 from Taunton via Chard Junction to Seaton was 6s 3d

Below: The basic platform shelter at Seaton with staff and a dog happily posing for the photographer. The reconstruction of the station by the Southern Railway during the 1930s was to sweep all of this away. *Lens of Sutton*

return, second class, the train departing from Taunton at 11am, arriving at Seaton at 1pm and departing for Taunton at 7.15pm — a good day out for a bargain price.

The original signalbox was situated at the entrance to the yard but alterations were to come, for on 22 July 1929 the Southern Railway's Traffic Officers Conference considered and recommended a proposal to remove the single-line tablet instruments etc from the signalbox to a new ground frame. This was to cost £572 but would result in an annual staff saving of £327. The proposal was approved by the Traffic Committee and put into action on 11 January 1930, the ground frame being situated near the stop blocks on Platform 1. After this date the signalbox was used only when the bay platform (No 2) was brought into use. On 27 May 1936 the crossover at the north end of the station connecting the two platform roads was taken out of use, thus allowing the platform to be extended and widened. The original signalbox was closed on 28 June 1936, when a completely new track layout came into operation during the remodelling of the station. A new signalbox was installed adjacent to the platform near the terminal end under the canopy, the box containing 20 levers, four of which were

spare, in a Westinghouse A2 frame. The original Italianate station was demolished and replaced by a new station predominantly constructed in concrete to an Art Deco design. The single platform was retained and lengthened to accommodate 12 coaches and covered for a length of 300ft, the eastern face being used for departures and arrivals, with the bay on the western face used for carriage storage. The run-round loop gave access to the new locomotive shed which was now situated at the terminal end of the station. A new coaling stage, water tower and cattle pens were also constructed in concrete, the only building to survive the rebuilding being the goods shed which received a cement rendering, becoming an integral part of the station itself.

With Seaton and its neighbouring resort of Beer being the two most popular holiday resorts in East Devon, the Southern Railway decided to market the town as a 'working class' resort, and this meant that thousands of passengers would have to be provided for at summer weekends. Better facilities would be necessary for the benefit of the travelling public, hence the alterations at Seaton and also, to a larger extent, at Seaton Junction. The rise of Seaton as a holiday resort led to a much improved summer

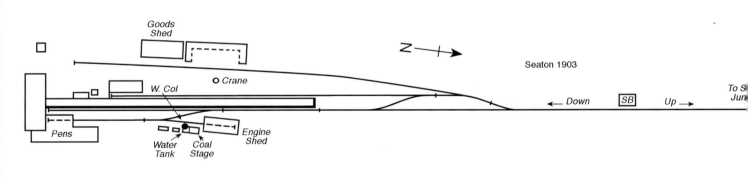

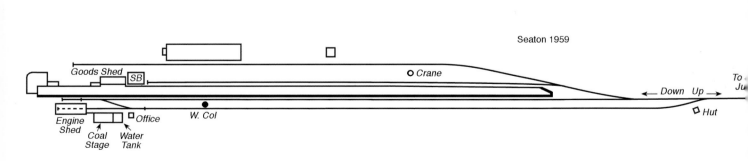

Above: The handsome Art Deco station of 1936 at Seaton, seen here on 29 September 1956, brought a touch of Southern suburbia to East Devon. *H. C. Casserley*

service, even more so when a Warner's holiday camp was opened in 1935 practically adjoining the station and bringing more trade to the railway. Holiday traffic was healthy at the station right up to the 1960s.

From 1 June to 30 September 1913 the daily through coaches to Seaton comprised a corridor tri-composite and a corridor brake third by the 1pm ex-Waterloo. The summer service for 1914 shows nine passenger trains running each way, including one down train which omitted the Colyford stop, a mixed train, and one up and down goods working. Branch services were first and third class, although second class had not yet been abolished generally on the South Western at that time. Excursion traffic, as expected, was considerable up to and including the 1930s on weekdays and Bank Holidays, with through trains operating from Waterloo, Exeter and Yeovil. The summer timetable of 1938 shows services at their peak, 13 passenger trains running each way on Mondays to Fridays, with an additional working on Thursdays and one of the services running through to Axminster. Journey times for passenger trains were 12 minutes (omitting the Colyford stop) or 16 minutes. Saturday services increased to 15 up and 14 down passenger trains, including the Axminster through service, with an additional locomotive being provided from Exmouth Junction to work three down and four up services. A freight train was the first working

service of the day, departing from Seaton at 6am with a return at 7.30am. Sunday trains comprised 14 up and 13 down services.

The same timetable for 1938 shows through coaches on the 11am from Waterloo despatched onwards from Salisbury by the 12.38pm slow on Mondays to Fridays, and in the opposite direction, by the 9am Ilfracombe-Salisbury from Seaton Junction, and thence by the 12.30pm from Exeter. On Saturdays there were through coaches on the 9.1am, 12.9pm and 3.24pm from Waterloo and, in the reverse direction, the 9.55am and 11.30am from Exeter, and the 2.12pm Seaton-Waterloo. As always with country branch railways from then on, progressive introduction and popularity of buses and cars affected passenger services and loadings, although there was a passenger increase during the late 1940s and early 1950s. Through carriages from and to Waterloo operated on weekdays, with complete through trains on summer Saturdays — the postwar services in October 1947 had 13 passenger trains running in each direction on weekdays and four on Sundays. The summer timetable of 1948 had a similar service on Mondays to Fridays, rising to 15 on Saturdays, plus six return journeys on Sundays increasing to seven in 1949. Through coaches ran on the 10.50am (later 11am) from Waterloo and the 9.35am from Exeter, the latter train conveying the coaches to Templecombe for collection by the following 10.30am up service.

Above: Pictured before the reconstruction of the station, 'O2' class 0-4-4T No E224 has uncoupled from the branch set at Seaton on 24 June 1928 and is reversing over the points, before running forward over the rails to the far left in order to enter the shed. *H. C. Casserley*

Two coaches were the norm on weekday branch services, to which an extra all-third was added on Saturdays and Bank Holidays. Sunday services, which had been started in the 1930s to combat bus competition, were reduced to one coach in the 1950s and withdrawn in the winter timetable of 1961-2, being replaced by buses from Axminster. The first carriages to be used on the branch at the opening were four-wheeled stock, and probably the oldest at that, giving way to six-wheeled vehicles circa 1910 which were used briefly until the arrival of the bogie push-and-pull sets. The LSWR built three new, gated two-coach push-and-pull sets in 1914, each 56ft in length, comprising an all-third and a composite driving trailer, with the first class compartments adjoining the guard's and luggage compartment behind the driver's control section. Constructed in similar design to the steam railmotors but of much heavier construction, they were originally fitted with the Drummond wire and pulley system, but were converted to the compressed air control system from 1928. Consequently, two of the new sets (SR set No 373 comprising third class No 738 and composite control trailer No 6545 and set 374 — Nos 739 and 6547)

were used in the district and one of the sets, usually No 373, ran regularly on the branch up until the end of the 1930s, reappearing occasionally in later years in between visits to the Bulford branch. Incidentally set No 373 also appeared on occasions on the Yeovil Town-Junction branch, and I well remember during my footplate days at Yeovil Town station, as also reported on the Seaton branch, that when a mass of passengers waiting to alight had crowded into the centre vestibule, only to discover the iron gates opened inwards, they had to retreat in order for the gates to open correctly! The situation was made worse if passengers waiting impatiently on the platform to climb aboard were already trying to force their way in.

The LSWR searched high and low for a suitable vehicle to strengthen the Seaton branch set during the holiday season and in 1916 unearthed a converted 'Eagle Express' saloon fitted with end gangways, this being used on the line until 1922-5 for third class conveyance only. The coach was 8ft 0¾in wide compared with the 8ft 6in of the regular branch set, and when the 'Eagle' saloon was placed in the centre of the branch train for the summer season, the appearance of the set was rather spoilt, although in

Above: Adams 'O2' No E224 stands alongside the water tank and coal stage for locomotive requirements at Seaton on 24 June 1928. *H. C. Casserley*

later years the saloon was placed at one end of the train. The 'Eagle' vestibule stock was introduced in 1893, formed into nine-coach sets for use on Southampton liner trains. Two short-bodied ex-LSWR sets, Nos 734 and 735, formed of composites Nos 4759 and 4760 and control vehicles 2644 and 2645 appeared briefly in the late 1940s. Originally constructed for use on the Channel Islands boat trains, they had been converted to push-and-pull operation in 1943. Push-and-pull set No 381 was converted in 1949 from the old steel-panelled LSWR stock with tapering ends to the brake vans (the 'Ironclads' introduced in 1921) and comprised control vehicle No 4052 and composite No 6560, both vehicles being 57ft long. Also to receive similar treatment was set No 385, with Nos 3213 and 6564 being put aside for working the branch as well as the Swanage and Lymington branches when converted in 1949.

Set No 723, an ex-LBSCR through set, appeared on the branch in the early 1950s, this vehicle being 54ft long with low arc roofs and comprising No 3855 (a six-compartment third brake control vehicle) and No 6250, a nine-compartment composite (two first and seven third). Both vehicles had been built for

push-and-pull operation c1921. An ex-SECR 60ft 10-compartment third, No 1066, was used for many years in the 1950s for strengthening the branch set at peak times; it was stabled at Seaton Junction for most of the time until 1958, when it became an integral part of set No 1 with No 6488 (replacing former LSWR third No 2620) and reappeared briefly in its new guise on the branch in 1961. A similar but steel-panelled vehicle, No 1098, took over as the Seaton strengthening vehicle in 1958, and this vehicle is now preserved on the Bluebell Railway. Another SECR strengthening vehicle appeared on the branch in the 1950s: No 1050 had originated from Lancing Works in 1927 and consisted of five compartments from an ex-SECR second class six-wheeler, three from a similar third class vehicle, and two compartments of uncertain origin, the whole mounted on a frame 62ft 6in in length as used for electric stock. Renumbered 5546, this composite ran in set No 760 on the Eastern section until 1943 when it was downgraded to all-third and fitted for push-and-pull working. It appeared at the Centenary Exhibition at Exeter Central and, after ending its days on the Swanage Branch, also survives on the Bluebell Railway.

Above: Adams Radial '0415' class No 3488 at Seaton with the 10.10am through carriages to Waterloo on 9 July 1949. The new engine shed, which opened on 28 March 1936, can be seen on the left. *H. C. Casserley*

British Railways converted Maunsell corridor stock into set No 616 comprising composite brake control vehicle No 6695 and third open No 1358 (until replaced by No 1359 in November 1961). Entering service in 1959, they appeared on the branch in 1960, as did set No 603 comprising composite brake control No 6675 and open second No 1320. These sets were the last steam conversions to push-and-pull working on BR, and they were also the last SR stock on the Seaton branch. Carriages were usually propelled to Seaton with the engine working bunker first. From 3 May 1963 products of that other railway, based in Swindon, appeared on the branch when '6400' class pannier tanks and auto trailers Nos 238 and 253 appeared; Nos 240 and 235 were also later used on the line. Although not participating in the branch workings, a departmental 12-wheeled coach of LNWR origin languished for a long while at Seaton, standing on an isolated piece of track beside the station and used by the motive power department. Another unusual vehicle was the goods brake used on the branch for many years. It was originally built as a cattle drovers' van and, prior to its use on the Seaton branch, was used on the Callington line to convey quarrymen.

Seaton Locomotive Shed

The first engine shed, as mentioned previously, opened with the line. Constructed of timber and measuring 45ft x 18ft it was deemed unsuitable even for the smallest locomotives such as the Beattie 2-2-2 well tanks. An inspection pit was provided inside the building, with a sleeper-built coal stage situated outside. The shed, which soldiered on for nearly 60 years, was described as alarmingly close to the waters of the River Axe — and creaking weirdly at times as if it was resisting a slide into the muddy water below. So cramped was the building that when the 'D1' 0-4-2Ts arrived, the footplate crews had only inches to spare when boarding the locomotive. The old shed was eventually demolished during the station reconstruction and replaced by a building constructed in concrete, having a corrugated pitched asbestos roof. The new shed was situated off the new lengthy run-round at the east side of the station. A coal stage, a water tank and column were provided immediately outside the entrance, and doors were provided at the

134

Above: A six-wheel LSWR cattle drovers' brake, No S54962, in brown livery with red ends stands in the yard at Seaton on 11 March 1952. It previously worked on the Bere Alston-Calstock-Callington branch in the 1930s where it was used to convey quarrymen. *A. E. West*

Below: The driving trailer of set 381 at Seaton on 12 March 1952 — an LSWR/SR 'Ironclad' brake third, No S4052S, in Malachite green livery. *A. E. West*

Above: With all the pipes and plumbing associated with a push-and-pull-fitted locomotive, 'M7' 0-4-4T No 30021 takes water at Seaton on 16 July 1958. The Westinghouse pump is bolted to the side of the smokebox and the large air tank can be seen under the front end. This locomotive, dating from January 1904, was fitted with air equipment in July 1930 and was withdrawn in March 1964. *A. E. West*

southern end of the shed enabling the dock it was built against to be used for end loading. The new construction opened on 28 March 1936, with motive power staff consisting of two pairs of drivers and firemen plus an engine cleaner for coaling. The two pairs of footplate crews worked one early and one late turn week and week about. A signing-on cabin was provided plus sleeping accommodation in the form of the old LNWR coach for the men who were lodging away from Exmouth Junction shed which was responsible for the Seaton shed. Locomotives were outstationed overnight in order to work the first up service in the morning. As mentioned previously, the Beattie 2-2-2 well tanks of the 'Tartar' class, Nos 12 *Jupiter* (dating from June 1852) and No 33 *Phoenix* (July 1852), are known to have been used on the branch in the 1860s. The locomotives, when new, were used in the London area, with *Phoenix* being outstationed at Northam by 1862. The 0-6-0 Beyer Peacock 'Ilfracombe Goods', Nos 282 and 284, were regular visitors to the line by mid-1885 — No 282 becoming 0282 in the duplicate list in 1889 and subsequently No 0349 in June 1900. Withdrawn in May 1909, the locomotive eventually entered Kent & East Sussex-Railway stock as No 7 *Rother*. Stablemate No 284 became 0284 in the duplicate list and also ended up on the K&ESR stock list as No 9

Juno. The 2-4-0 Beattie standard well tanks were also used on the branch in the latter years of the 19th century, with No 182 allocated in March 1878 and the March 1890 listings showing No 198 in use on the line. Adams 'T1' and 'O2' 0-4-4Ts were also used at this time, with 'O2' 0-4-4T No 213 becoming the branch engine until transferred away in mid-1914, being replaced in June of that year by two more members of the class, Nos 201 and 204. Both locomotives had been modified with the Drummond cable and pulley system of motor control. The small Drummond 'S14' class 0-4-0 motor tanks Nos 101 and 147 were used on the branch in 1911, but were found to be unsuccessful as the coal consumption was little better than an Adams 'O2'. It is from 1913 onwards that the Lyme Regis 4-4-2 tanks (when spare) were to be seen on the line, and motor-fitted (cable and pulley) No 0486 in the early years of the grouping relieved the 'O2' until withdrawn in 1928. Adams 'O2s' Nos 183 and 236 were used on the branch in the mid-1920s, also using the cable and pulley system until events elsewhere (see Yeovil chapter) in 1929-30 forced the Southern Railway to forbid its use.

An ex-LBSCR Stroudley 'D1' 0-4-2T, No B234, performed a successful trial on the line on 29 January 1930; the panting of the Westinghouse pump

Above: Part of the signalbox can be seen on the platform to the right as 'M7' No 30021 prepares to leave Seaton with the 3.17pm to Seaton Junction on 24 August 1958. *Terry Gough*

reverberating around Seaton Junction from this class of engine (and the forthcoming 'M7s') was to set the scene for over 30 years. On 19 June of the same year the 'D1s' started work on the branch. Fitted with the ex-LBSCR system of push-and-pull control operated by compressed air, they ran with Brighton two-coach gated saloon sets converted to air control which was more practical and safer than the LSWR system. The 'D1s' working the push-pull services in 1932 comprised Nos B214 and B256. No B214 was originally No 13 of the LBSCR dating from 1874 and, except for a Marsh boiler and injectors, remained almost as originally constructed when allocated to Seaton. For many years prior to electrification, and running as No 347, this locomotive was based at Dorking shed working on the demanding and heavy London suburban services. Also operating at Seaton at the same time was Adams 'O2' No E182. Having been fitted with the Brighton system of compressed air control for push-and-pull operation in November 1932 and September 1933 respectively, 'O2s' Nos 187, 183 and 207 replaced the 'D1' locomotives on the branch duties, only to be replaced themselves by similarly fitted 'M7' 0-4-4Ts, with No 45 allocated to Exmouth Junction for Seaton branch duties in October 1932. The 'M7s' remained in almost complete charge of the branch services from then on almost to closure.

Another 'M7', No 27, was used on the line on 14 July 1936, and Nos 27, 46 and 55 were recorded as being on the line in mid-1937, with Nos 46, 49, 55 and 105 at work in the closing years of the Southern Railway. Nationalisation brought new numbers to some old faces, with 'M7s' Nos 30021, 30045, 30046, 30048, 30105 and 30480 in use until 2 May 1963 when No 30048 and a three-coach set was used on the branch for the final time, being replaced by ex-GWR '6400' 0-6-0PT No 6400 and auto-coaches. However, after a fortnight's duty, on 18 May 1963, the pannier tank disappeared and an Ivatt 2-6-2T, No 41272, took over the branch duties and, as this particular locomotive was not fitted with push-and-pull apparatus, it had to run around the Swindon-built trailers at each end of the journey — not very helpful when turnaround times were tight or when trains were running late. The pannier tank reappeared a few days later but unfortunately failed again, being replaced by Ivatt 2-6-2T No 41309 which stayed on the branch for a week until the errant pannier returned on 10 June. Other panniers used on the line were No 6430, with 6412 arriving in July of the same year.

In pre-1914 days, the Adams Radial 4-4-2 tanks were employed working many excursions on Bank Holidays, and in 1948 one of the class worked the Seaton branch regularly during the summer, arriving

Above: An ex-LSWR 58ft composite (rebuilt by the SR), No S4617S, part of set No 35, is pictured at Seaton on 16 July 1958. *A. E. West*

each Saturday at approximately 8.30am to pilot the branch tank on the 9am up train which consisted of 10 coaches. The Radial then returned to Seaton light engine, before hauling by itself the 10.10am nonstop to Seaton Junction. This was probably the last working on which one of this class of locomotive was regularly booked to pass a station. The working continued for a few years, including working in the afternoon through portions of down trains, until replaced by another 'M7' on this duty early in the 1950s. Upon the withdrawal of restrictions in 1960, SR Moguls and Bulleid Light Pacifics were allowed on the branch. 'N' class Moguls were employed on through trains, and as there were no turntables between Yeovil Junction and Exeter, this was not the most popular trip for footplate crews as it involved a fair bit of running tender first. The unrebuilt Bulleid Pacifics also appeared on such working, and in 1962 hauled the Saturdays only through train to Waterloo, the locomotive arriving at Seaton at about 7.15am with a single coal truck plus a goods brake. 'U' class locomotives seen on the branch in 1963 included Nos 31792 and 31798, and also 'West Country' class No 34030 *Watersmeet*.

In May 1963, an early morning goods trip to Seaton was usually worked by a BR Standard 2-6-2T, with the locomotive returning light engine to Seaton Junction and the branch tank working the return goods. During the severe frosts of 1963 the water column at Seaton was out of order, the nearest operational one being at Axminster, but unfortunately the working timetable did not allow for the branch locomotive to run there. A call for help to Exmouth Junction produced a Bulleid Light Pacific to take over the branch set while the 'M7' was being refilled by hosepipe from the station toilet at Seaton! After one hour, enough water had been taken aboard for the tank engine to make a dash for Axminster for a complete refill. The Light Pacific was on the branch for two days before it was released back to Exmouth Junction.

Diesel multiple-units first appeared on the line on 4 November 1963. The new motive power was not an unqualified success, with BR Standard Class 3MT 2-6-2T No 82040 and a Mk 1 brake second carriage having to stand in for a failed DMU on 26 September 1964. Other diesel failures led to ex-GWR '1400' class 0-4-2 tank locomotives Nos 1442 and 1450 being sent down from Yeovil Town shed to Seaton on 7 February 1965, with No 1442 in use on the branch and No 1450 held as a standby engine. The DMUs returned to the line after a month's absence and both of the '1400' class tanks were placed in store on

Above: Well laden with passengers, car No 2 of the Seaton & District Electric Tramway prepares to leave Colyford en route to Colyton on 6 July 1993. The level crossing can be seen to the right. This car was built in 1964 as an open-top double-decker based on a Metropolitan Electric Tramways/London United Tramways design. *Hugh Ballantyne*

11 May 1965 and were saved from extinction, with No 1442 preserved at Tiverton and No 1450 initially on the Dart Valley Railway. Locomotive diagrams used for freight and passenger duties on the branch in 1963 included Exmouth Junction Duty Nos 606 and 607.

The line flourished well into the late 1950s. Two coaches were the norm for the branch train on weekdays for the summer traffic, and an all-third strengthening coach was added on Saturdays and Bank Holidays, the branch train having to start with seven, or on at least one occasion eight, coaches on the 1 in 76 from the platform end at Colyton. This era witnessed some of the heaviest trains seen on the branch, with 10 or more coaches being handled by the branch engine assisted by an Adams Radial 4-4-2T on some Saturdays, and the longest train noted on one Sunday in September 1961 was a monster of 11 bogie luggage vans and a brake second. Not requiring to stop at Colyton, it ran through with a tremendous roar from the exhausts of both locomotives. However, this was not good enough for the upper strata of the railway management system, with the Western Region gaining control of all former Southern Region lines west of Salisbury on 1 January 1963. More dark clouds lay on the horizon with the Beeching Report recommending closure. Goods traffic ceased on the

branch on 3 February 1964; the signalbox closed on 2 May 1965 and, except for the line into Platform 2, all of the trackwork was taken out of use and the tablet working replaced by 'one engine in steam' working. Complete closure was effected on 7 March 1966, with Driver Harold Pope, who had worked on the branch for many years, driving the last diesel multiple-unit on the final day. Public support on the last day was not encouraging — only about a dozen people attended. Seaton Junction also closed to passengers on the same day. Seaton station was demolished in 1969, the site levelled, and is now occupied by an electronics factory. The concrete bridge constructed across the River Axe by the Seaton & Beer Railway in 1877 is still standing, although a modern replacement now stands alongside; the former toll house also remains.

Rail closures meant that many people who had worked most of their lives and given loyal service on the railways were thrown out of their jobs. The farcical situation at Axminster station in the summer of the closure of the Seaton branch reflects the then attitude to staff and passengers alike by an uncaring management. The station was besieged by travellers detraining at the station and travelling to the holiday camps at Seaton by bus. When the hourly bus duly

Above: The fireman of 'M7' No 30045 concentrates on the job while engaged in shunting movements in the yard at Seaton on 12 March 1952. *A. E. West*

arrived, most of the passengers could not get on board, and there was certainly no room for the vast amount of luggage. They either had to wait another hour for the next bus or pay their own taxi fare. The local press likened the scene as akin to 'Napoleon's retreat from Moscow'. The following year witnessed railway staff being forced into the ridiculous situation of having to book buses and coaches to and from Axminster for holidaymakers travelling to Seaton which the now-closed branch line could have catered for quite adequately.

A relatively happy ending to the closure of the Seaton branch was sounded by the arrival of the Seaton Tramway constructed to a gauge of 2ft 9in, and which now operates on three miles of trackbed. Services start from the car park off Harbour Road at Seaton, where a magnificent Victorian style terminus surmounted by a fine clock has been constructed. A depot to house the trams has been built on the former station throat, and it is from here onwards that the original trackbed to Colyford and Colyton is used. Electric tramcars, fed from overhead wires, now trundle along the route once worked by the Beattie well tanks and the 'M7s'. Colyford station is still used and has a passing loop and a down bay. The cast-iron gents' is still in place, but is stranded high on part of the original formation as the platforms are now set lower to suit the trams. The level crossing over the main road is equipped with barriers and traffic lights;

the tram driver enters his key, presses a plunger, traffic lights flash amber then red to stop the road traffic, the barriers are lowered, and the tram then moves across the road. The wheel flanges operate a treadle which then cancels the red lights, allowing the barrier to be lifted. Various pieces of the former railway can be seen as the trams travel along the route. Colyton is the terminus of the tramway operations; the bulk of the platform exists, as do the station buildings, which are now converted into two private houses. The small store building at the up end of the platform is still standing, the former goods shed is in situ, as are the goods offices. A tea room and gift shop exists at the station where passengers using the tramway can relax and enjoy a good brew and a slice of cake before exploring Colyton, one of the most delightful and charming towns in East Devon with its interesting shops set amongst quaint streets only a 10-minute walk from the station, before returning to Seaton. It is interesting that the tramway now brings visitors to Colyton from the nearby holiday camps — which is exactly what the former railway used to do.

Footplate Memories

The 8-mile-long ascent from Seaton Junction from the valley of the Axe to the valley of the Otter was considerable for down trains, being a test of man and

Class 4MT 2-6-4T No 80041 at Seaton with an LCGB special on 7 March 1965. *S. C. Nash*

Above: Wreathed in steam, which seems to be escaping from most parts of the locomotive, ex-GWR No 1450 is in a poor condition as it stands alongside Platform 1 at Seaton on 20 February 1965. *S. C. Nash*

machine in the days of steam, and like many footplatemen I have had my share of trips, good and bad, in fair weather or foul with Bulleid Pacifics, 'Arthurs', 'S15s', 'U' and 'N' classes and 'T9s'. One of my favourite class of engine, especially for freight work, was the workhorse of the Southern — the 'S15' 4-6-0 or 'Blackuns' as they were nicknamed. We used to have a diagram that involved working a heavy freight from Yeovil Junction to the main freight yard at Exmouth Junction and, as was often the case with the Southern at that time during the late 1950s, the train would be fully loaded with coal, timber, petrol, box vans, etc. After a fairly good run we would arrive at Axminster where we would take water, drop off and pick up wagons from the yard. As usual on a dark winter morning, it would be raining buckets with the wind howling around the cab. The sheet would have been the first thing to have been put up at Yeovil, the heat of the fire being reflected under its protective cover.

While we take water, now is the time to drag the best of the coal forward for the climb ahead. My mate is outside putting some oil around the straps and glands; smoke is billowing around the station from our chimney, with steam drifting everywhere. Levers thud and bells ring in the box as the signalman prepares our way to Seaton Junction. The water bag is

taken from the tender, the flap shut with a clang, and we are ready to go. The guard has already given us the load and is now in his van waiting for the off. Ahead of us green aspects twinkle in the darkness and the cosy glow from the signalbox windows reflects against our long boiler, the heat from the firebox drying off our wet clothes, despite water dripping down through the holes in the tender sheet. With plenty of steam showing on the pressure gauge we are ready; water is dancing up and down in the gauge glasses; the gauge lamp gives a warm glow to the brass and copper pipes on the boiler. With a long shriek on the whistle, my mate carefully opens the regulator and promptly closes it again; the compression comes off the buffers on our wagons and we ease forward. More regulator, but not too much in case we break a coupling. With a thunderous beat from our chimney, the engine gets into her stride as we pull underneath the Lyme Regis branch overbridge. The firebox door is closed now so as not to let cold air be drawn on to the tubeplate. A glance over my side of the footplate catches the guard's lamp swinging, indicating that the train is complete. I answer with one of the spare headlamps which is always lit and kept on the footplate during the hours of darkness. Equipped with a red shade, as well as the white aspect, it is always ready just in case of a derailment, or worse, on the main line. Our locomotive is straining away. With the reverser wound back and the regulator wide open, every piece of a steam locomotive is alive when she is pulling hard, and there is no sensation in the world to equal that of a locomotive footplate; that is, if everything is going all right, as you were never sure when the next rough trip was going to happen.

Time now to put a few rounds in the firebox. As the firedoor is opened, you squint your eyes to avoid being temporarily blinded by the white roaring mass. The heat is welcome as the cab is a draughty place to work. Turn from the tender flap, into the box where you want it to go, turn again and place another shovelfull in the box. The blast on the fire is so great that most of the coal is sucked from the shovel by the power of our exhaust. Enough firing for a bit, as we are nearing Seaton Junction; the dawn is just breaking, but the rain is still coming down in sheets. A rearward glance sees our long train rumbling and rolling behind us, and the white side lamps on the guard's van at the rear can be seen, indicating that our train has not snatched a coupling. My mate has the collar of his railway-issue mackintosh pulled up and his cap jammed down over his ears. He gazes through the cab window trying for that first glimpse of the Seaton Junction distant signal; the vacuum gauge is rock steady at 21 inches. His hand is never far from the vacuum brake handle, just in case something may go wrong. All footplatemen, myself included, always listen to the locomotive; she will tell you if something is wrong, maybe it is something in her beat, maybe with her cylinders, or the smell of a hot bearing, but you always know when something is not quite right. The water is coming down in the gauge glass. I open the water valve and the steam valve of the injector, and as I stick my head out of the cab to check that the injector is picking up at the overflow, the force of the storm hits me fair and square and nearly wrenches my waterproof cap from my head.

There is a loud ring from the AWS system bell, and with the Seaton Junction distant signal now showing a green aspect, at the very least we will now have a fair go at the bank. We rattle through the station with steam beating down on our long train, and we're now past the advanced starter and hitting into the bank. No banker to assist us, as would be the case on the Western Region. We are on our own, and it is up to us to reach the top. The 'S15' is rock steady for the first mile and then the train begins to hold on to our tail; the injector is switched off, steam pressure is holding. Now for the furnace as shovel after shovel of coal is fed into the firebox under the door. This is how the 'Blackuns' like it. My mate works the firedoor, opening and closing as I feed the fiery furnace, at the same time keeping a wary eye on the road ahead. Higher and higher we climb the bank, each length of rail that we pass over is a small victory. Our exhaust is almost vertical, with smoke and steam screaming out from the chimney; the cab is pulsating with power distributed from the massive forces at work in the

cylinders. The 'S15s' are very surefooted and well liked by all of the crews, especially when used on freight traffic, although they are just as good on passenger turns. They are fairly free steaming engines, if fired correctly, and don't gobble half the amount of coal that a Bulleid Pacific does.

With our speed down to almost a crawl, and the reverser nearly in full forward gear, we are barking up the bank, the locomotive still plodding away. A glance rearwards to our long train shows the sheets of rain pounding against the wagons. After another bout of firing, it's time for a quick brush up of the footplate, removing any stray lumps of coal. On with the injector again, and time for a quick Woodbine before another round of firing. My mate refills his pipe and has a quick glance at the gauges, noting we have adequate supplies of steam and water in the boiler. There is no time to relax for at any moment a gauge glass could burst or the engine could slip on a greasy rail. The closed firedoor rattles in time with each mighty beat of the exhaust, the heat under the cab sheet is almost suffocating, but it is better than having a wet back. Puddles of water form in the tender and now and again when filling the firing shovel, streams of black slurry pour on to the footplate. The Honiton Incline box down distant arm is off, a welcome sight as this means that the tunnel mouth is not far away, although not a time to feel pleased with ourselves as we are now at the steepest part of the bank. There is a roar and a whistle as an up train speeds down the incline. We climb past the signalbox. The signalman with his hands on the levers in readiness to slam the boards back to danger behind us has already asked for the road to his counterpart at Honiton, and signalmen in boxes all down the line as far as Exmouth Junction are already aware of our progress and are preparing the road ahead for our train. The blast of the exhaust rebounds around the cutting as our train painfully drags itself into the gaping tunnel mouth; a long scream on our whistle echoes inside the tunnel. The exhaust slams against the roof; sparks and char hurl down on to the locomotive and the leading wagons. The rumbling of many wheels grows louder and louder as our long train is dragged further and further into the black confines of the interior. Steam and smoke almost suffocate us; the blinding white glare from the firebox is reflected against the cab and the tunnel walls. No more firing now as we blast our way through the tunnel. Just up ahead is a small pinprick of light which grows larger and larger as we near the exit and as we do so the miracle of Honiton Tunnel emerges, for all of a sudden our exhaust beat softens. We are now running downhill. The regulator is eased, the reverser is wound back as we gather speed with our massive train pushing us along. Now is the time to top the boiler up, have another Woodbine, relax and let the fire look after itself for a bit as we speed towards Honiton and our destination at Exmouth Junction.

All in a morning's work — the footplate; there was nothing like it.

*The timings shown on this page **will not apply on Saturdays, 23rd July to 10th September,** inclusive.*

EXETER & SIDMOUTH JUNCT. (VIA EXMOUTH) AND SIDMOUTH BRANCHES.

Distance from Exeter.		WEEK-DAYS.	Pass.		6.10 a.m. Freight. Exm'th Jc. Sidings. ★		Pass.		Pass.				Pass.		Pass.		Pass.			
M.	C.		arr.	dep.	arr.	dep.	arr.	dep.	arr.	dep.			arr.	dep.	arr.	dep.	arr.	dep.		
			a.m.	a.m.	a.m.	a.m.	a.m.	a.m.	a.m.	a.m.			a.m.	a.m.	a.m.	a.m.	a.m.	a.m.		
...	...	Exeter	...	5 55	...	...	...	...	...	7 15	...	...	...	...	...	8 15	...	...	...	...
...	41	Lion's Holt Halt.........	...	...	...	...	...	...	...	...	...	...	...	...	...	...	...	...		
1	14	Exmouth Junction	...	5 59	6 11	6 12	...	...	...	7 19	...	...	...	...	8	19	...	...		
1	39	Polsloe Bridge Halt	...	•••	...	...	...	...	...	7 22	...	...	...	...	...	8 22	...	...		
3	19	Clyst S.M. & Digby H.	...	...	...	...	...	...	...	...	...	...	...	...	...	...	...	...		
5	34	Topsham..........	6 7	6 8	6X21	7 0	...	...	7 30	7 32	...	...	...	...	8 30	8 31	...	...		
6	14	Odam's Siding	...	...	...	...	...	...	...	...	...	...	...	...	...	...	...	...		
6	75	Woodbury Road	...	6 12	...	...	...	...	...	7 36	...	...	...	...	...	8 35	...	...		
8	37	Lympstone	...	6 16	7	7	...	...	...	7 40	...	...	...	...	...	8 39	...	...		
10	48	Exmouth	6 21	...	7 12	...	...	6 40	7 45	...	...	...	...	8 8	8 44	...	...	...		
12	18	Littleham	...	...	...	...	6 45	6 46	...	...	...	8 13	8 17	...	...	...	...		
15	33	Budleigh Salterton	...	...	...	...	6 54	6 57	...	...	...	8X25	8 27	...	...	...	...		
17	44	East Budleigh	...	...	...	...	7 1	7 2	...	...	...	8 31	8 32	...	...	...	...		
18	63	Colaton Raleigh Sg.	...	...	...	...	...	...	...	...	...	...	...	...	...	...	...		
20	45	Newton Poppleford......	...	...	...	...	7 7	7 8	...	...	...	8 37	8 38	...	...	...	...		
21	67	Tipton St. John's	...	...	...	...	7 12	...	...	...	...	8 42	...	...	...	...	...		
M.	C.																			
3	16	**Sidmouth**	...	...	...	...	...	7 8	...	...	...	...	...	...	...	...	8 38		
...	...	Tipton St. John's	...	...	...	...	7 17	7 19	...	...	...	...	...	...	...	8 47	8X49		
23	77	Ottery St. Mary	...	...	...	...	7X23	7 24	...	...	...	...	...	...	...	8 53	8 54		
26	74	**Sidmouth Junction**	...	...	...	...	7 32	...	...	...	...	...	...	...	...	9 2	...		

WEEK-DAYS.	Pass. C		Pass. C		Pass.		Freight.		Pass. Sidmouth.		Pass. Exeter.		9.18 a.m. Freight Exmouth Jct. Sds.		Pass.			
	arr.	dep.	arr.	dep	arr.	dep.	arr.	dep.	arr	dep.	arr.	dep.	arr.	dep.	arr.	dep.		
	a.m.	a.m.	a.m.	a.m.	a.m.	a.m.	a.m.	a.m.	a.m.	a.m.	a.m.	a.m.	a.m.	a.m.	a.m.	a.m.		
Exeter........	...	...	...	...	...	8 47	...	...	...	...	...	...	...	...	...	9 40	...	...
Lion's Holt Halt	...	...	...	...	...	...	...	...	...	...	...	...	...	...	...	9 42½	...	...
Exmouth Junction	...	...	...	...	8 51	...	...	...	...	...	...	9 19	9 20	9 45	...	...	...	
Polsloe Bridge Halt	...	...	...	...	...	8 53	...	...	...	...	...	...	...	...	...	9 47	...	...
Clyst. S.M. & Digby H..............	...	...	...	...	...	8 58	...	...	...	...	...	...	...	...	...	9 52	...	...
Topsham	...	...	...	...	9 3	9 4	...	...	...	...	...	9X29	11●10	9 57	9 59	...	...	
Odam's Siding	...	...	...	...	...	...	...	...	...	...	...	11 13	11 18	...	...	...	...
Woodbury Road	...	...	...	...	...	9 8	...	...	...	...	...	11 21	11 29	...	10 3	...	...
Lympstone	...	...	...	...	...	9 12	...	...	...	...	...	11 34	11 44	...	10 7	...	...
Exmouth	...	...	9 5	...	9 17	...	...	...	...	9 45	...	11 50	...	10 12	...	...	...	
Littleham	...	...	9 10	9 11	...	...	...	...	9 50	9 51	...	...	...	...	...	...	...
Budleigh Salterton	...	...	9X 19	9 22	...	...	...	...	9X 59	10 2	...	...	...	...	...	...	...
East Budleigh	...	...	9 26	9 27	...	...	...	...	10 6	10 7	...	...	...	...	...	...	...
Colaton Raleigh Sg.	...	...	...	...	...	...	...	...	...	...	...	...	...	...	...	...	...
Newton Poppleford.............	...	...	9 32	9 33	...	...	...	...	10 12	10 13	...	...	...	...	...	...	...
Tipton St. John's	...	...	9 37	...	...	...	...	...	10 17	10 35	...	...	...	...	...	...	...
Sidmouth	...	9 22	...	...	...	...	...	9 58	...	...	...	10 17	...	...	...	...	...	
Tipton St. John's	9 31	...	X	9 42	...	...	10 10	...	...	...	10 26	10 30	...	...	...	...	...
Ottery St. Mary	...	...	9X 46	9 47	...	...	...	...	...	...	10 34	10 36	...	...	...	...	...
Sidmouth Junction	...	...	9 55	C	...	...	...	...	...	...	10 44	10 48	...	...	...	...	...

C—Through coaches for L.M.S. Rly., via Templecombe, attached on Fridays from 22nd July to 9th September, inclusive, departing Sidmouth Junction 10.2 a.m.

Sidmouth Junction-Sidmouth and Tipton St John's-Budleigh Salterton-Exmouth

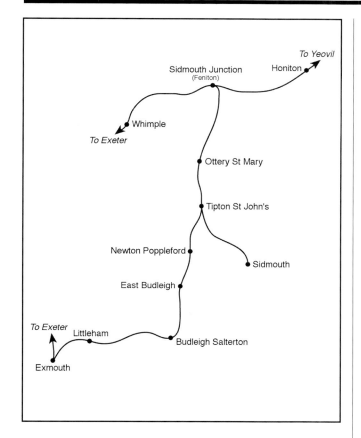

The Sidmouth Railway

Sidmouth was a fashionable and attractive health resort long before the arrival of the railway, and was considered a 'genteel' seaside town to visit. Like several Devon resorts developed during the Napoleonic wars when the wealthy and elite could not reach the Continent, the town was also enjoyed as a winter retreat due to its mild climate, with a theatre being established as early as 1805. The Duke and Duchess of Kent stayed there in 1819 at Woolbrook Cottage with their infant daughter, who was later to become Queen Victoria. Even today, much of the town with its Regency and early Victorian buildings has a quiet residential character and charm all of its own, not possessing the 'brasher' air of some of the livelier South Devon resorts. Remaining select even to this day, caravan sites are noticeable by their absence, so very different from Seaton which was

promoted by the Southern Railway as a 'working class' resort. However, trade, such as it was, was diminishing early in the 19th century but two proposed harbour projects in 1811 and 1825 were of no avail. Another similar plan in 1836 involved a narrow gauge railway which was constructed to convey large blocks of stone to the esplanade from a reef of rocks approximately 1½ miles to the east. A steam locomotive was delivered by sea for the railway, but the lack of facilities meant that it could not be unloaded at Sidmouth, and the ship was sent to Exmouth for unloading, the locomotive being drawn by horses thence to Sidmouth and placed upon the railway. Unfortunately being found to be too large to pass through the tunnel which had been constructed through Salcombe Cliffs, the locomotive was then used for entertainment by carrying the townsfolk along the esplanade; good for the townsfolk but of no use to the investors who lost £12,000 out of a working capital of £15,000. The locomotive had gone by 1838 and the tunnel was sealed up. One of the bricked-up portals can still be seen today.

During the railway mania Sidmouth, as with many other areas, found itself involved in one way or another with schemes which may have been a good idea at the time. The following two plans proved fruitless: a branch to Sidmouth was in the proposals of the Exeter, Yeovil & Dorchester Railway in 1846, while the Devon & Dorset Railway, a joint venture by the Great Western and the Bristol & Exeter Railway for a line from Maiden Newton to Exeter with a branch to Sidmouth, was proposed in 1852. A direct rail communication to London was badly needed, as the only rail route to the capital at the time was by road transport to Exeter and thence onwards via Bristol and Swindon.

With the main line station at Feniton opening on 18 July 1860, rail communication was now only nine miles from the resort. A meeting was held at Sidmouth on 18 December 1861 for the purpose of planning a railway from Sidmouth to Feniton, and combining this with a harbour scheme. As a result, the Sidmouth Railway & Harbour Co received its Act of Parliament on 7 August 1862, authorising the raising of capital of £120,000 and £40,000 on loan, of which £90,000 would apply to the railway. The

LSWR was authorised to lease the line for an annual sum of £5,000. However, money was slow to arrive, the scheme proving to be unpopular with subscribers. Another Act, dated 5 July 1865 authorised the company to build a half-mile-long line at Sidmouth for which no additional capital was to be raised; unfortunately it was found at a meeting held on 30 November 1866 that the company had liabilities of £20,000 of which £16,000 was owed to the railway contractor. The company collapsed in 1869 with only some of the line's earthworks completed. Revival of the scheme for the railway (the harbour had now been forgotten) occurred with a further Act of 29 June 1861 receiving the Royal Assent; this had been promoted by the trustees of the Balfour family, who had purchased the Manor of Sidmouth. The directors of the Sidmouth Railway were Sir John Kennaway MP (Chairman); John Fulford Vicary of North Tawton; John Macmillan Dunlop of Windermere; Neil Bannatyne of London; and John Heugh of London. With an authorised capital of £66,000, and borrowing powers of £22,000 this was a sound proposition and construction went ahead.

The agreement of 17 March 1871 between the directors of the Sidmouth Railway and the LSWR stated that the line could be constructed as a light railway in accordance with the Regulation of Railways Act of 1868. The Sidmouth Company was to maintain the line for one year from the opening and the LSWR would work it for 55% when gross earnings did not exceed £4,000, and 50% when above that figure. The South Western also had the option to purchase the railway. In July 1872, the Sidmouth Board accepted the tender for the sum of £35,000 offered by R. T. Relfe of Okehampton as contractor. The first three of the 'Ilfracombe Goods' 0-6-0s, Nos 282, 283 and 284, entered service in March 1873 and, after being run in from Nine Elms, were then loaned to the engineers department for ballasting between Sidmouth Junction and Sidmouth. The line was laid with Vignoles pattern 60lb/yd rail fixed with fang bolts and spikes to transverse sleepers. Although land had been purchased to make a double line, the branch was laid as a single line with loop lines and sidings at the stations; four level crossings, equipped with gates interlocking with the signals, were provided over public roads. Engineering works included seven over- and seven underbridges and two viaducts. Colonel F. H. Rich RE made the Board of Trade inspection on 2 July 1874 and, apart from a few matters that were drawn to the attention of the engineer, sanctioned the opening of the Sidmouth Railway. Stations on the line included Ottery St Mary, Tipton and Sidmouth, with turntables at Sidmouth and Ottery Road. The line opened for traffic on 6 July 1874, although it would be fair to say that since the resort of Sidmouth had been attracting many fashionable visitors for 80 years, especially for winter residence before the arrival of

the railway, the opening of the line made less difference here than at any other seaside resort in the West Country. The Sidmouth Railway Company, despite an offer from the LSWR in 1894 to purchase the line for £70,500, remained independent until 1923.

The Budleigh Salterton Railway

The intention of the Sidmouth, Budleigh Salterton & Exmouth Railway to build from a junction at Tipton St John's on the Sidmouth line was put forward in 1862. However, ideas were revised and the Sidmouth & Budleigh Salterton Act of 28 July 1863 authorised the construction of a line from Tipton as far as Budleigh Salterton, with the LSWR working the line. Not making progress with this scheme, a similar plan was put forward, with a deviation in the Harpford-Colaton Raleigh district, resulting in the Sidmouth & Budleigh Salterton Act of 5 July 1865. This proposal also ended in failure. A bold attempt was then made to construct an extension from Sidmouth, with the Sidmouth Railway Act of 11 August 1876. The LSWR was to have worked the line; the agreement for this dating from 19 May of the same year. The proposed line was to start from Sidmouth station, where a junction was to be made with the recently opened Sidmouth Railway and, running parallel to the coast, a 475-yard-long tunnel was to be bored through Peak Hill. The railway would then cross the River Otter near East Budleigh, making a junction with the Exeter & Exmouth Railway near its terminus. Unfortunately this grand scheme failed to realise its potential when the Sidmouth Company found itself unable to raise the authorised capital of £130,000, the abandonment of the lines authorised in the 1876 Act being made law by the Sidmouth Railway Act of 23 May 1879.

The people of Budleigh Salterton were determined though to have rail communication. A plan to build a line from Tipton St John's station to a terminus midway between Budleigh Salterton and Otterton Point in 1886, however, also failed in Parliament. A similar venture was promoted in 1893, the chief opponent being the Exmouth Dock Company, who proclaimed that the amount of dock income received by goods carried from Exmouth to Budleigh Salterton was approximately £500, and if they were to lose this, then their net income would fall to £1,500 per annum. All was to no avail, as the townspeople now had the bit between their teeth and were determined to have their railway, despite the opposition of the dock company. The Act of Incorporation for the Budleigh Salterton Railway was passed on 20 July 1894, authorising the company to construct a line from Tipton to Budleigh, the first directors being the Hon Mark Rolle, Dr Robert Walker and Dr Thomas Brushfield. The

LSWR was to work the line, taking £65 for junction costs at Tipton St John's and 60% of the balance of gross receipts yearly, paying the remainder to the Budleigh company, plus the sum of £150 for office expenses. Messrs Lucas & Aird were the appointed contractors, who employed 400 navvies on the project. The ceremony of cutting the first sod was performed at Greenway Lane, Budleigh Salterton by Lady Gertrude Rolle, wife of the landowner through whose land the new line would traverse. Delay was caused in 1896 when the River Otter burst its banks and washed away two temporary bridges erected by the contractors. However, the contract was finished six months ahead of time and Major F. Marindin RE inspected the line for the Board of Trade on 10 April 1897. The line was not constructed to light railway standard, as had been the case with the Sidmouth line, and used double-headed steel rails of 82lb per sq yd laid on sleepers of creosoted Memel timber. Signals were not provided, except at Salterton, and all points connected with the main line were worked from ground frames locked by the train staff. The line was sanctioned to open with stations on the line comprising Budleigh and Salterton, and siding connections at Newton Poppleford and Colaton Raleigh. The line opened to public traffic on 15 May 1897, and was the last passenger branch line to be opened in Devon before the end of the century. An agreement to purchase the line by the LSWR was authorised by an Act of 18 August 1911, with the take-over being effected on 1 January 1912.

Exmouth & Salterton Railway

Another link in the chain of railway communication for East Devon began in November 1896, with the trustees of the Rolle Estate proposing that the line be extended from Budleigh Salterton to Exmouth. The original plan ran into trouble from the start, with objections from the Budleigh Salterton Railway and also the citizens of Exmouth. The new line was planned to run from Exmouth station via Imperial Road and the Manor House grounds, through a tunnel under the Beacon and eventually travelling through the Littleham Valley and the cliffs to Budleigh. The inhabitants of Exmouth would not have their town divided by a railway, with a petition to the Hon Mark Rolle making no doubt as to their opposition. This now forced the engineers to mark out the more difficult and expensive option passing through Marlpool Park to Littleham Cross. The eventual cost of the line, when built, was £111,378, the engineering structures now demanding deep cuttings and viaducts. Authority for the building of the 4 mile 46.3 chain Exmouth & Salterton Railway was contained in the LSWR Act of 25 July 1898. The contractors, Henry Lovatt & Sons, had considerable experience

with constructing sections of the Great Central Railway and the Manchester Ship Canal, and construction was under the surveillance of J. W. Jacomb-Hood of the LSWR assisted by the resident engineer, E. Roach.

The new line was to extend from the north end of Exmouth station and terminate by making an end-on connection with the former terminal station at Budleigh Salterton. One of the major engineering features was the 30ft-high curving brick-built viaduct with a length of 352 yards approaching the junction with the Exeter-Exmouth line and consisting of 23 arches incorporating two girder bridges which was built to accommodate double track although always laid to single. A brick-built three-arch bridge carried the Salterton road over the line. The cutting at Knowle caused problems, and the soil excavated from here was taken to the foreshore near Exmouth station, eventually providing a site upon which the future Exmouth goods depot was developed. Two steam navvies plus an industrial steam locomotive were employed on the works, and by June 1901 work on the deep cutting in Marlpool Park had begun with construction of the embankment on the south side of Dalditch Lane proceeding with vigour. The steam navvy at the same location was advancing in earnest towards Knowle Hill, with its companion working from the Exmouth side of the hill. The LSWR had considered plans for making a triangular junction outside Exmouth which would have allowed through running from Exeter Queen Street to Budleigh Salterton. However, the scheme was dropped due to the land required being very expensive. Major J. W. Pringle made the necessary inspection of the line for the Board of Trade on 26 May 1903. One intermediate station was provided at Littleham, with two platforms and a passing loop. At Exmouth station the island bay platform was extended and, with an additional set of facing points provided along with new signalling and trackwork for the junction, the signal cabin was extended and the lever frame relocked. The new extension opened to traffic on Whit Monday, 1 June 1903.

Upon leaving Sidmouth Junction the line passed under the A30 trunk road and, after descending a gradient of 1 in 110, ascended at 1 in 53, the line beyond being relatively easy as it followed the valley of the River Otter through Gosford Gates Crossing and Cadhay Gates Crossing to Ottery St Mary and Tipton St John's. From Tipton the Sidmouth line climbed at 1 in 45 for two miles before reaching the summit at Bowd, and then descended rapidly at 1 in 54 before arriving at the terminus situated approximately one mile from the beach and 200ft above sea level. Trains leaving the junction at Tipton St John's for Exmouth had a falling gradient of 1 in 50, which eased to 1 in 360 before crossing the River Otter and proceeding along the right-hand bank to Newton Poppleford. A level stretch of line beyond Newton Poppleford station soon changed to a descent of 1 in 100 before easing to 1 in 217. The Otter was

Above: Sidmouth Junction, looking towards the level crossing in 1930. Sidmouth trains departed from the opposite face of the down platform on the left, from which a graceful canopy provided shelter for passengers. No such facility existed on the up platform in the foreground, except for a large brick-built waiting shelter which can be seen at the far end of the platform. *Lens of Sutton*

crossed again, with the line levelling to 1 in 940 and then ascending at 1 in 290 to cross the River Otter once more. The line passed Colaton Raleigh Siding and then descended at 1 in 290, easing further upon arrival at East Budleigh. Leaving here, the line climbed for 1½ miles on a gradient of 1 in 50 and, crossing over a 54ft-high brick-built arch spanning Dalditch Lane, the summit of the line was reached at the deep Knowle Cutting, before descending at 1 in 50 for ¾ mile before Littleham station. From here the line descended at 1 in 50 and, passing under the brick-built bridge carrying the Salterton road above the line,

wended its way through the outskirts of Exmouth, skirting Phear Park before crossing over Exmouth Viaduct and finally joining the line from Exeter before entering the station.

Sidmouth Junction

Opened to public traffic as Feniton on 19 July 1860 and situated one mile from the village of the same name, the station was renamed at various times before reaching junction status, becoming Ottery

Road the following year on 1 July, then Ottery St Mary in April 1868, and eventually becoming Sidmouth Junction when the branch line to Sidmouth opened on 6 July 1874. The first branch train was seen off by approximately 200 people. The wheel then turned full circle, with the station reopening as Feniton on 3 May 1971, four years after closure. The main station buildings, of brick construction, were situated on the down platform, containing the station offices, two waiting rooms (one for ladies), and the booking office. It was constructed to the usual Tite design. The down platform and branch bay shared a long canopy, and a large brick-built waiting shelter was provided on the up platform. W. H. Smith maintained a small bookstall under the platform canopy here for many years. Both platforms were connected by a footbridge. A level crossing was (and still is with the new Feniton) situated at the west end

of the station and, in steam days, the gates were operated from a small gate box containing five levers positioned on the Broad Clyst side of the crossing. A staff of 20 was based at the station, including a stationmaster, booking clerks, shunters, porters, crossing keepers and signalmen. The distinctive signalbox dating from 1875 was positioned east of the station alongside the up siding and up main line, and a member of the station staff had the task of walking to the box to collect or return the single-line tablet for the branch trains, the tablet being lowered from the box on a long rope. Extensive siding accommodation was provided at the junction, and a turntable was positioned in the fork created by the main and branch lines, but this was removed in 1930. The goods shed contained a 40cwt crane and stood alongside the bay platform line, while a 5-ton crane was to be found in the down goods yard which handled bricks, coal, stone, timber, lime and agricultural machinery, with outward bound traffic including milk, potatoes, sugar beet and cider apples.

The Sidmouth trains arrived and departed from a bay at the rear of the down platform. An engine release road was not provided, and this meant that arriving trains had to be reversed out in order to allow the engine to run round. Seven daily passenger trains hauled by the Standard Beattie 2-4-0 well tanks ran each way when the branch to Sidmouth opened in 1874. No Sunday services were provided and, from November of the same year, the early morning service was curtailed, leaving a pattern of six trains per day. Trains to Sidmouth were designated as down, and towards Sidmouth Junction as up. Adams 'O2s' were at work on the line in the 1890s and, with the opening of the Budleigh Salterton line extension, a number of Waterloo-Exmouth through services were routed via Sidmouth Junction and Tipton St John's from as early as 1914. A service from Nottingham via Bath and the S&D ran from 1927, and on Saturdays in 1932 the 11.10am and 3.10pm arrived from Waterloo as well as the Nottingham train. The all-Pullman 'Devon Belle' instituted a new service on 16 June 1947. Departing from Waterloo at 12 noon, it arrived at Sidmouth Junction at 3.16pm (to connect with the Sidmouth and Budleigh Salterton) and departed at 3.20pm for its 16-minute dash to Exeter. After its suspension during World War 2, the 'Atlantic Coast Express' began running again on 6 October 1947, departing from Waterloo at 10.50am, arriving at Sidmouth Junction at 2.8pm where the Sidmouth and Exmouth coaches were detached. The through trains made Sidmouth Junction a very busy place at times; the locomotives of such trains destined for Exmouth or Sidmouth arriving from Waterloo would then travel as light engines to Exmouth Junction and, during the 1960s, it would be usual to view six tank locomotives standing in the sidings at around 10.30am. Working in pairs, the locomotives had hauled the 9.25am Sidmouth-Waterloo and the 9.25am Exmouth-Waterloo, another locomotive had worked the 9.38am Littleham-Waterloo, and the remaining one the 10.17am from

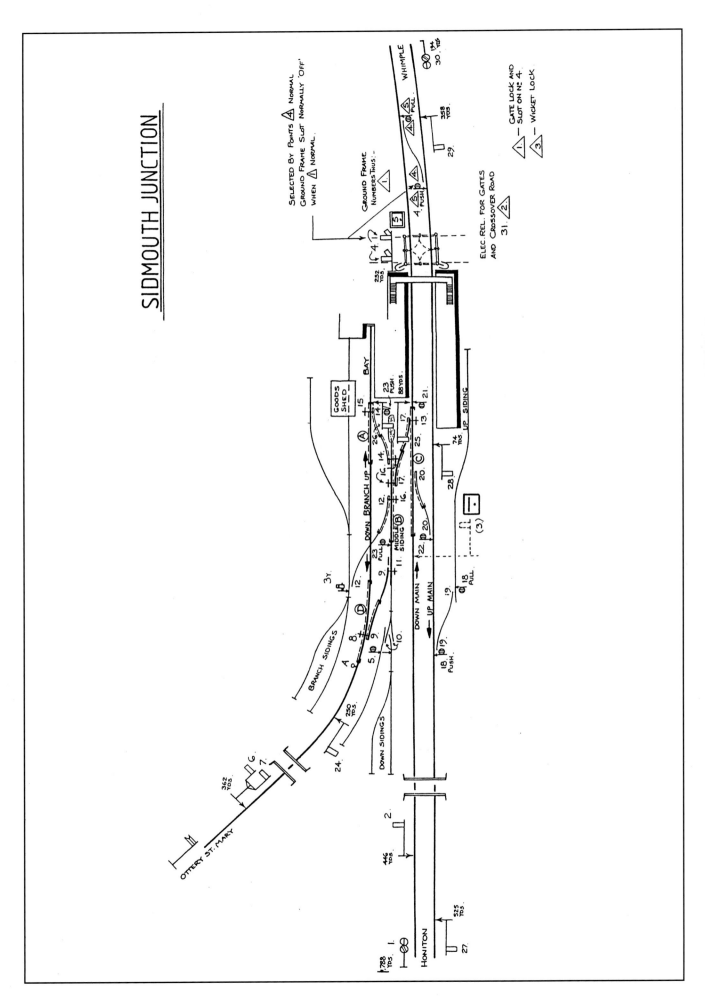

SIDMOUTH JUNCTION

Above: Maunsell 'S15' 4-6-0 No E831, then less than a year old and allocated to Salisbury shed, has reversed alongside the down platform at Sidmouth Junction on 4 August 1928 in order for mails and parcel traffic to be unloaded. Smoke deflectors were fitted in December of the following year. No 831, dating from September 1927, was withdrawn in November 1963. The coaches of a Sidmouth train can be seen in the branch bay to the left.
H. C. Casserley

Sidmouth. The Monday to Friday through coaches were withdrawn in 1964. Sidmouth Junction closed to goods traffic on 6 September 1965, closing entirely from 6 March 1967 with the cessation of passenger services to Sidmouth and from Tipton St John's to Exmouth. The signalbox also closed on 21 May of the same year after the final branch freight workings. However, due to strong local pressure, the station was reopened on 3 May 1971 under its original name of Feniton, but just before the reopening, the down platform was cut to half its original length and the Tite station building destroyed. The level crossing gates were replaced by barriers in 1974, and today the station has a good level of commuter passenger traffic to and from Exeter Central.

Gosford Gates

Gosford Gates level crossing was situated midway between Sidmouth Junction and Ottery St Mary. A five-lever Westinghouse open frame, installed on 23 March 1921, operated the gate lock, up distant, up home, down distant and down home. The Cadhay Gates down distant signal shared the same post as the Gosford up distant. This was not a block post and the gates were worked by hand while a relay bell was installed in the crossing keeper's cottage. Coal, water and other supplies were conveyed to the crossing keeper by the 9.10am freight from Sidmouth Junction.

Cadhay Gates

Cadhay Gates was equipped with exactly the same type of frame as its neighbour half a mile away. The five-lever open frame operated the gate lock, up distant and up home signals, and the down distant and down home. The Gosford Gates up distant shared the same post as the Cadhay down distant; the Ottery St Mary down distant was on the same post as the

Left: The Sidmouth bay starting signal at Sidmouth Junction, formed of an SR upper quadrant arm affixed to an LSWR lattice post, clears the way ahead for a branch line service. A main line service for Waterloo departs in the background, behind a Bulleid Pacific on 8 September 1961. *A. E. West*

Right: The fireman of 2-6-2T No 82023 relaxing on the barrow with his newspaper before working the 3.15pm to Sidmouth from Sidmouth Junction on 10 July 1958 looks suspiciously like my old mate, Dave Fling, who was a fireman at Exmouth before transferring to Yeovil Town. The builder standing on his ladder against the goods shed seems unconcerned by the noise generated by the lifting safety valves of the locomotive. *Terry Gough*

Cadhay down home, while the Ottery St Mary up advance starter shared the same post as the Cadhay up distant. The gates were worked by hand, with a relay bell installed in the crossing keeper's cottage. A porter from Ottery St Mary was responsible for servicing the gates although, in later years, churns of water were set down by a light engine or goods train.

Ottery St Mary

Ottery St Mary station was situated in the Otter Valley about half a mile west of the attractive town centre famous as the birthplace of the poet Samuel Taylor Coleridge (1772-1834), his father, the Reverend John

Above: Gosford Gates ground frame and crossing keeper's cottage on 6 October 1964. *A. E. West*

Below: The five-lever ground frame at Gosford Gates, 6 October 1964. Lever No 1 operated the gate lock, No 2 the up distant, No 3 up home, No 4 down home and No 5 the down distant. *A. E. West*

Above: Cadhay Gates level crossing and cottage, looking towards Sidmouth Junction in October 1964. The up home signal stands in the foreground with the down home and the Ottery St Mary down distant signals in the background. The ground frame and signals were brought into use on 23 March 1921. *A. E. West*

Below: The ground frame at Cadhay Gates in October 1964. The coal stove to keep the crossing keeper warm can be seen in the hut. Two cans of drinking water are by the small wooden platform used for such deliveries and perhaps the odd bag of coal etc from the locomotive footplate. *A. E. West*

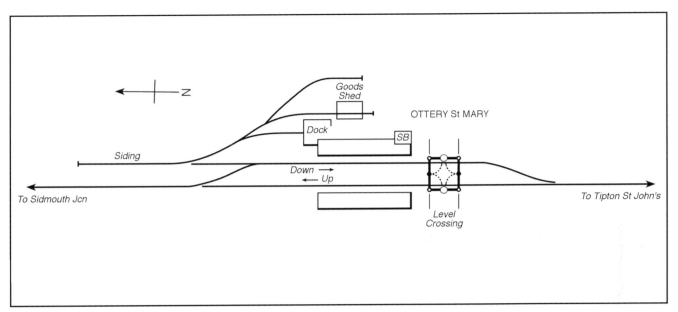

Above: The main brick-built station buildings at Ottery St Mary were situated on the down platform, seen here on 6 October 1964. A passing loop was provided which was extended in 1936 to deal with longer trains on the summer services. *A. E. West*

Above: An extensive brick-built waiting shelter was provided on the up platform at Ottery St Mary, viewed here in October 1964. *A. E. West*

Below: Ottery St Mary station as viewed from the approach road on 21 September 1963. The goods shed is on the right with a solitary box van standing in the siding. The building to the far right bears the name of the well-known local firm of Miller & Lilley. *H. C. Casserley*

Above: Ottery St Mary goods shed on 6 October 1964 with its small office to the left of the sliding door. *A. E. West*

Below: A side view of the goods shed at Ottery St Mary showing the large sliding wooden doors at the entrance used by road vehicles. Part of the down platform can be seen in the foreground. *A. E. West*

Above: Ottery St Mary signalbox, down starter and level crossing on 20 September 1961. The signalbox opened on 20 November 1955, replacing the former box dating from 1894 which stood opposite at the end of the up platform. *A. E. West*

Left: The up home signal at Ottery St Mary consisting of an upper quadrant arm on a post made from two rails bolted together. *A. E. West*

Coleridge being the vicar of the town. The station was provided with a crossing loop, which was extended southwards for two chains on 22 November 1936. Three sidings and a goods shed were situated on the down side of the layout. The main station building consisted of a two-storey gabled, brick-built house which was located on the down platform, with a brick-built waiting shelter on the up platform. A small timber-built ground level signalbox, installed in 1874, stood off the Tipton St John's end of the up platform near the level crossing, this being replaced on 20 November 1955 by a modern brick-built cabin situated on the opposite side of the track. This new box remained in use for only just over 10 years. The level crossing gates over the B3174 road were worked by a gate wheel from the box. The station in later years was well used by the children attending King's School, Ottery St Mary, who travelled on the 8.55am from Sidmouth. Cattle trains were run from the station when the Ottery St Mary market was held on alternate Mondays and, more than once, if the load was too great for the normal goods train, then a

Above: A charming view of the level crossing at Tipton St John's in the early 1900s. The Golden Lion hostelry stands to the right just beyond the crossing. The young girl in her long dress and the cottages bathed in summer sunshine, devoid of road traffic, reflect an era of English rural life that has gone for ever. *Lens of Sutton*

special would run from Ottery St Mary to Sidmouth Junction. Coal traffic was handled here until 8 May 1967, exactly two months after passenger services were withdrawn. The former station building and goods shed still survive. Two miles further on from the station the River Otter was crossed by a 55-yard-long viaduct, which is also still in situ.

Tipton St John's

Opened as Tipton, the station served the village of the same name situated to the east, was originally only a passing point on the Sidmouth line. It was renamed Tipton St John's on 1 February 1881. The main station building was on the up platform, with a timber-built (later altered to brick-built) waiting shelter on the down platform. The original signalbox was located on the Sidmouth side of the level crossing, but was replaced in March 1897 when the station achieved junction status with the Budleigh Salterton Railway which opened on 15 May of that year. The new box, sited at the Sidmouth end of the down platform, was equipped with 32 levers and a gate wheel. A four-lever ground frame was also provided at the up end of the station alongside the down loop until its removal in February 1930, its former function being taken over by the signalbox. A footbridge connecting the two

platforms came into use in February 1898, replacing a sleeper crossing. It is interesting to note that the Sidmouth line was known as the main, and the Budleigh loop as the branch. With the opening of the line to Budleigh Salterton, and eventually to Exmouth, the station became a very busy junction between two single lines, especially on summer Saturdays when long through trains from Waterloo were divided, and joined, with as many as 50 trains calling here. Use of the East Devon branches was encouraged by the introduction just before World War 1 of runabout tickets, a system used by day trippers from London from 1900. Camping coaches were accommodated on one of the sidings behind the up station buildings. A large water tank stood on the Exmouth side of the level crossing alongside the up line.

The steep incline of the line to Sidmouth contrasted with the Exmouth line, which had an easy falling gradient to about half a mile beyond East Budleigh and could be seen to good effect from the end of the platform. The train services at the station were considerable, making this location one of the busiest junctions between two single lines to be found anywhere in the country, and all dealt with in the style that was only to be found on the Southern Railway and its successor, the Southern Region. Through trains rolled in to be split or divided and

160

Above: A passenger's-eye view of the level crossing and junction at Tipton St John's on 7 July 1962. With the Sidmouth branch dead ahead climbing at 1 in 45 for nearly one and a half miles, the line to Budleigh Salterton and Exmouth swings away to the right. The footbridge was constructed in 1898 after the station became a junction.
A. E. West

sent onwards again, with shunters, footplatemen, station staff and signalmen all performing, at times, some very complicated diagrams with professionalism second to none. The 1938 weekday summer timetable had 11 passenger trains running each way between Sidmouth Junction and Sidmouth, two Sidmouth Junction-Exmouth, two Ottery St Mary-Exmouth, and five Tipton St John's-Sidmouth workings, three of which ran through to and from Exmouth, plus the workings from Tipton to Ottery St Mary. Freight workings were also catered for, with the 7am Sidmouth Junction-Sidmouth, and the 7.45am Exmouth Junction-Exmouth, returning at 2.35pm and 10.46am respectively. The Exmouth passenger workings had connecting services between Tipton and Sidmouth. Three passenger trains each way ran through to or from Exeter, being the 5.50am, 7.34am and 8am from Exeter Central and the 10.20am, 3.15pm and 10.25pm from Sidmouth. The same 1938 timetable shows down through coaches on Mondays to Fridays to Sidmouth and Exmouth on the 11am and 3pm from Waterloo, and to Sidmouth only on the 1pm from Waterloo, while on Saturdays there were also coaches on the 8.38am, 9.1am, 12 noon and 3pm trains. In addition, there was the 10.24am Derby to Sidmouth and Exmouth running via the Somerset & Dorset and Templecombe. The weekday up services had through

coaches attached to the 10.30am and 12.45pm from Exeter Central, and on Saturdays the up service was increased, with carriages being added to the 9.55am and 11.5am from Exeter, plus a 2.20pm ex-Sidmouth. The Derby through coaches (a service which did not reappear after World War 2) were put on to the 9.30am from Exeter Central.

The service on summer Saturdays in 1938 consisted of 16 down and 15 up services between Sidmouth Junction and Sidmouth, two between Ottery St Mary and Exmouth, and six Sidmouth Junction-Exmouth, plus three down and two up Tipton St John's-Sidmouth. Goods workings and through workings to Exeter also ran. The Sunday services had 11 down and 12 up trains, plus a return working between Sidmouth and Exmouth, and also an Axminster-Sidmouth service. The winter 1938-9 timetable to Sidmouth was almost identical to the Monday-Friday summer service, consisting of 23 trains each way. The summer of 1953 saw three Saturday through trains from Waterloo to Exmouth and Sidmouth, all of which had to be reversed at Sidmouth Junction and double-headed to Tipton St John's where, upon arrival, they were divided. The 8.5am ex-Waterloo consisted of seven carriages, the two parts being hauled beyond Tipton by their respective engines. The 9am from Waterloo was a different matter, however, as it was formed of 12 coaches, the seven-coach Exmouth

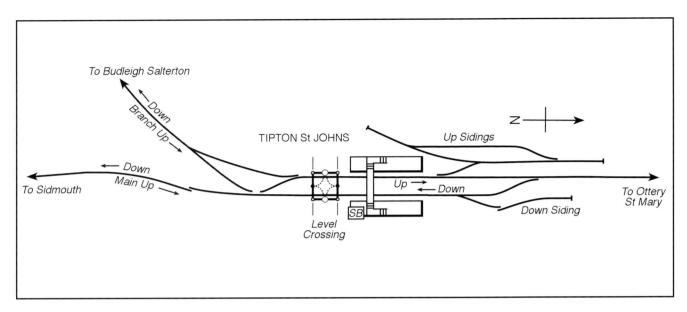

portion being taken onwards by two engines, leaving the remainder to be worked to Sidmouth by another. The 11.45am ex-Waterloo consisted of eight coaches for Exmouth and four for Sidmouth. The 7am ex-Cleethorpes was rerouted from its original destination of Bournemouth to Sidmouth and Exmouth in 1960. This is described on Table 42a of the 1962 Western Region, British Railways summer timetable as a 'Through Train Service from Cleethorpes, Grimsby, Lincoln, Nottingham, Leicester & Birmingham to Axminster (for Lyme Regis), Seaton Junction (for Seaton), Sidmouth, Budleigh Salterton and Exmouth'. The train ran from Cleethorpes via Grimsby Town, Market Rasen, Lincoln (St Marks), Nottingham Midland, Loughborough, Leicester London Road, Birmingham New Street, Gloucester Eastgate, Bath Green Park and the S&D to Templecombe, thence down the Southern main line to Axminster arriving at

4.21pm, going forward to Seaton Junction (4.29pm), Sidmouth Junction (4.51pm) and Tipton St John's (5.5pm). Here the train was split into portions, giving arrivals at Sidmouth (5.17pm), Budleigh Salterton (5.26pm) and Exmouth at 5.40pm. What a relief it must have been for the passengers as they disembarked, having been on the train since 7am! The return working comprised the 10.42am ex-Exmouth and 11.07am ex-Sidmouth, both sections joining at Tipton St John's, with the stock comprising ex-LNER Gresley and Thompson carriages alternating with SR stock. This remarkable through service ceased to run from 1 September 1962 due to the withdrawal of long distance trains from the Somerset & Dorset. The main station building survives in private ownership, complete with canopy, and a portion of the up platform also remains; however, houses have been built on the trackbed of the former junction.

Left: Class 2MT 2-6-2T No 84020 leaves Tipton St John's for Sidmouth on 20 September 1961. The train comprises a two-coach set formed of a Maunsell open third and a Maunsell brake compo. The Bulleid brake compo at the rear is a through coach from Waterloo. *A. E. West*

Above: Tipton St John's, looking towards Ottery St Mary in the 1930s, with the main station buildings situated on the up platform. The up starter is an LSWR lower quadrant. The running-in board instructs passengers to 'Change for East Budleigh; Budleigh Salterton and Exmouth'. *Lens of Sutton*

Below: Pullman holiday coach No P48 at Tipton St John's on 8 September 1961. This coach, originally named *Sunbeam*, dated from 1921 and must have been sheer opulence compared with the old LSWR camping coaches located at various other locations along the line. *A. E. West*

Above: Class 3MT 2-6-2T No 82024 pulls away from Tipton St John's on 24 August with the 4pm Sidmouth Junction to Sidmouth. *Terry Gough*

Below: A pair of Drummond 'M7' tanks, Nos 30025 and 30024, pass Tipton St John's with a Waterloo to Exmouth special on 2 September 1962. *S. C. Nash*

Above: Ivatt 2-6-2T No 41318 leaves Tipton St John's with the 12.38pm to Sidmouth on 9 August 1960. *Terry Gough*

Below: Drummond 'M7' 0-4-4T No 30323 at Tipton St John's with the 2.4pm Sidmouth Junction to Sidmouth on 3 July 1956. *Hugh Ballantyne*

Above: The LCGB 'East Devon' rail tour departs from Tipton St John's bound for Exmouth hauled by 2-6-2T No 41206 and '5700' class 0-6-0PT No 4666 on 28 February 1965. *Hugh Ballantyne*

Below: The signalbox and level crossing at Tipton St John's on 24 May 1963. Dating from March 1897, the 32-lever box replaced the original signalbox which was situated on the Sidmouth side of the level crossing. *A. E. West*

Above: Drummond 'M7' 0-4-4T No S34 tackles the 1 in 45 from Tipton St John's with the 10.40am Exeter Central to Sidmouth on 15 June 1949. The 'S' prefix was carried by No 34 from 20 March 1948 until 4 April 1952, it then becoming No 30034. The locomotive, dating from April 1898, was withdrawn in February 1963. *S. C. Nash*

Below: The fireman of 'M7' 0-4-4T No 534 relaxes on the footplate while travelling through Harpford Woods with the 1.8pm Sidmouth Junction to Sidmouth on 15 June 1949. *S. C. Nash*

Above: Equine transport awaits potential customers at Sidmouth in the early 1900s. The larger carriage seems to have monopolised the space by the station canopy while the smaller gigs with their drivers and horses take shelter from the sun in the shade of the trees on the left. The large building to the right is the stationmaster's house. *Lens of Sutton*

Sidmouth

The station was situated almost a mile inland on the north-west outskirts of the town. The pleasing and attractive terminus consisted of a brick-built structure containing the stationmaster's accommodation and station offices. The station entrance had a wooden canopy above two arched doorways leading to the canopied island platform, the canopy being supported by decorative cast-iron brackets supported on cast-iron pillars. The eastern side (No 1 road) could accommodate seven coaches and the western (No 2 road) five, the platforms being flanked on the western side by the locomotive shed and turntable, and on the other by a substantial goods shed beside which there was another siding containing coal staithes for the local traders. Engine release from the No 1 road was effected by crossovers through the down siding, the No 2 road not having any engine release facilities, and the procedure for releasing a locomotive involved reversing the train out, running the locomotive into an adjacent siding (invariably the old shed road) and

then releasing the coaching stock back into the platform by gravity. A tall signalbox standing at the Tipton St John's end of the station was equipped with a Stevens frame containing 23 levers of which two were spare. The branch was at first worked by staff and ticket under absolute block regulations until replaced, in 1904, by the Tyers No 3 tablet, this being used on the main line until 1 January 1953 when this was superseded by the Tyers No 6 tablet. The original signalbox, of which no details appear to exist, was located near the locomotive coaling stage, the later box opening in 1905, probably when tablet working was introduced. A siding to the gasworks ran behind the signalbox, this being provided in the 1930s. Prior to then, coal for the gasworks was unloaded in the goods yard and taken by horse-drawn carts. Until closure of the locomotive shed, the engine was stabled overnight and the first train in the morning started from Sidmouth, the final train in the evening terminating there. The Southern Railway practice of holding the naming ceremony of the Bulleid 'West Country' class Pacifics at their relevant stations resulted in No 21C110 *Sidmouth* being named at the station on 27 June 1946.

1934 Southern Railway Working Timetable.

A goods train run between Tipton St John's and Sidmouth must have at the rear, a heavy brake van of not less than 20 tons, which, whenever possible, should be a van fitted with sanding apparatus. Should, however, a brake van of this description not be obtainable, two smaller brake vans, with a man in each, must be provided at the rear.'

Above: An Adams 'O2' 0-4-4T arrives at Sidmouth in the early 1900s. Lower quadrant arms can be seen on the gantry at the far end of the platform. A box van stands outside the goods shed to the right. *Lens of Sutton*

Below: Sidmouth in the early 1900s, with an Adams 'T1' 0-4-4T standing alongside the coaling stage in front of the turntable and locomotive shed. At least 11 goods wagons stand in the goods yard. *Lens of Sutton*

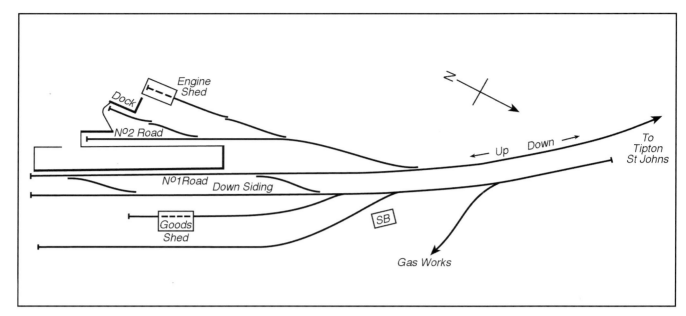

Train services to Sidmouth were considerable. Seven weekday trains ran each way in August 1887, increasing to eight down passenger trains, plus one mixed and a separate goods train, and in the reverse direction, nine passenger trains, one goods, plus a conditional goods from 1 October 1903 to 31 May 1904. The services then increased from year to year, and from 1 June to 30 September 1909 comprised 10 down passenger trains, plus one mixed, and one mixed from Tipton St John's; up trains consisting of 11 passenger plus a goods train. In the summer timetable for 1914, the services were now running at 11 passenger trains each way. As was to be expected with branches in the early days, coaching stock included four-wheelers, but by the early years of the 20th

century these had been replaced by six-wheeled stock. Set No 208, known to have worked in the area, comprised two brake thirds flanking an all-third and a tri-composite. Later, bogie stock appeared on Sidmouth branch services, including the familiar LSWR two-coach sets. The regular sets, both six-wheeled and bogie, were supplemented by odd vehicles for strengthening purposes, and there is evidence that some of these worked only between Sidmouth and Tipton St John's. From 1936 the sets consisted of a 61ft 7in LSWR rebuilt brake third on an SR underframe and an LSWR brake composite, and in sets Nos 42-46, both coaches were 61ft 7in rebuilds. The non-corridor coaches were eventually withdrawn and replaced by two-coach sets usually consisting of

Below: A view of Sidmouth station taken before World War 1. In pre-1914 summers, the LSWR operated three or four excursion trains each week from Waterloo to Sidmouth in addition to the regular through coaches. *Lens of Sutton*

Above: Adams 'T1' class 0-4-4T No 80 awaits departure from Sidmouth with the 6.15pm to Sidmouth Junction on 4 August 1918. *H. C. Casserley*

an LSWR brake composite plus an LSWR brake third, but various formations were put together, including brake thirds and brake composites, a brake third and a rebuilt composite, or a brake third plus a Maunsell composite. Through coaches from Waterloo over the years have been formed of a corridor brake third and brake compo, and a 48ft brake lav, tri-compo and a 48ft lav tri-compo.

In pre-1914 summers, the LSWR operated three or four 'excursion trains' each week from Waterloo to Sidmouth in addition to the regular through coaches. These trains were formed of five or six bogie thirds with a six-wheeled brake at each end. Loadings of passenger trains between Tipton St John's and Sidmouth were for the 'O2' class 128 tons, 'T1' class 150 tons and 'M7' class 160 tons; between Sidmouth and Tipton St John's: 'O2' class 150 tons, 'T1' class 170 tons and 'M7' class 180 tons. The summer timetable for 1964 had 11 down and 10 up trains on Mondays to Fridays, plus one each way between Sidmouth and Exmouth, this level of service increasing on Saturdays to 19 down and 21 up, plus two each way between Sidmouth and Exmouth, with the Sunday service having 10 trains in each direction.

Branch Telephones

Read minute of Traffic Committee of 8 June approving of telephonic communication being provide in the Western District as follows: Sidmouth Jct-Sidmouth (at a cost to Sidmouth Ry Co) £99 … To be carried out.

1714, 22 June 1898

Ref. To minute of 22 June, read letter from Sidmouth Company of 13 Oct objecting to bear the expense of telephonic communication upon their railway. Refer back to Traffic Committee.2107, 26 Oct 1898

2107, 26 Oct 1898

Ref. To minute of 26 Oct 1898, read minute of Traffic Committee 15 March stating that the Sidmouth Railway Co has now undertaken to bear the expense not exceeding £99 of providing telephonic communication on this branch railway.

624, 12 Apr 1899

Above: The exterior of Sidmouth station, looking towards the former locomotive shed on 25 June 1964. The wooden valanced canopy on the right was positioned above two semi-circular arched doorways leading to the canopied island platform. The parked cars are a reminder of the times and include a new Triumph Herald. *A. E. West*

Below: Another exterior view of the attractive station at Sidmouth taken on 25 June 1964. The final passenger train to run was the 6.57pm Sidmouth Junction to Sidmouth on 4 March 1967. *A. E. West*

The matter of telephone communication was one of some correspondence between the LSWR and the Sidmouth Railway Company, as the above extracts from the LSWR minutes of the engineering committee show. Another communication, regarding the gas supply for Sidmouth station, looks like a good deal on Mr Dunning's behalf, but as usual the LSWR would have come out best!

Gas Supply, Sidmouth...

Eng. Committee... ref. an offer from Mr Dunning to supply gas at Sidmouth station at the price of 5/- per 1000 cubic feet on condition that a coal depot be provided for him at that station free of charge. Approved and referred to engineering committee.

348, 7 Dec 1881

Above: A superb view of Sidmouth station taken from the starting signal gantry on 13 October 1959. The goods shed and yard are on the left, with empty coal wagons and a brake van standing outside. The canopied island platform has a train at the far end of the main platform while the bay line to the right is devoid of stock. The former locomotive shed on the far right is complete with water column and the remnants of the former coal stage. The turntable was removed for scrap in the 1920s and, being seldom used, the shed was closed in the mid-1930s. This was the only sub shed of Exmouth Junction to have been closed under the jurisdiction of the Southern Railway. *R. C. Riley*

Inward freight for the local traders, including Messrs Miller & Lilley, included coal (also for the gasworks), building materials, cement and timber. The closure of the gasworks brought about new traffic in the shape of Volkswagen vans for conversion into 'Caravanettes' by a local company, Devon Conversions. The vans were delivered to the former gasworks sidings by rail from Ramsgate. Coal traffic continued until 8 May 1967, but all general freight had been withdrawn on 6 September 1965. The station building is still extant and is used by a building company, the goods yard is now owned by Messrs Bradfords, and the station house is a private dwelling.

Sidmouth locomotive shed

The original locomotive shed was constructed in wood and measured approximately 45ft x 18ft. An inspection pit was provided inside, and a 42ft turntable was positioned immediately in front of the building along with a coaling stage, a small yard lamp and a water column. The water was fed by gravity and was the cause of some concern, with the Sidmouth Water Co offering to improve the service in 1897 at

Above left: The starting signals for the bay and main platform at Sidmouth on 24 March 1950. The LSWR wooden posts with decorative scrolls and spiked finials originally carried lower quadrant arms but these were replaced by upper quadrants in the 1950s. A Drummond 'M7' stands near the signalbox in the background. *A. E. West*

Above right: The rear of the Sidmouth starting signals on 25 June 1964. The signal arms themselves are of different eras. The one on the left is of Southern Railway parentage and the other arm is a BR version. A Ford Thames Trader BR delivery lorry stands in front of the goods shed. *A. E. West*

Left: Sidmouth signalbox was equipped with a Stevens frame containing 23 levers of which two were spare. The box closed on 8 May 1967. *A. E. West*

rates which would reduce the cost from £60 as paid, to approximately £30 per annum. Two sets of footplatemen were based at the shed, plus an engine cleaner who remained on duty overnight. The branch engine was usually changed over with a fresh locomotive at Sidmouth Junction once a fortnight, and although shedded at Sidmouth, its daily working took it to Exeter and Exmouth, with engines from other sheds working passenger trains over the branch during the course of the day. A severe fire destroyed the shed in the early hours of 7 January 1900, the origin of which is unknown, and it is reported that a number of supporting columns were the only remains to greet the staff the next morning. The sole occupant, an 'O2' 0-4-4T No 195, had the paint burnt off, but otherwise no major harm was done. The staff prepared her for the next day's working, and returned the locomotive to Exmouth Junction on the Monday evening, where it eventually received a new coat of paint. The shed was rebuilt in brick using the foundations of the original, but the new construction was not enlarged as it housed only a single tank engine sent out from the parent shed

Right: The former locomotive shed at Sidmouth in June 1964 showing the track cut short after closure of the shed and a buffer stop installed in front of the former entrance. *A. E. West*

Below: Sidmouth goods shed and goods office in June 1964. Closed to general goods traffic on 6 September 1965, the yard remained open to coal traffic until 8 May 1967. *A. E. West*

Above: The entrance to the goods yard and shed at Sidmouth in 1964. A rake of empty coal wagons has been marshalled in readiness to be collected by the local goods. *Lens of Sutton*

Below: A 1950s view of Sidmouth with an Ivatt tank on a single coach alongside the main platform and a rake of coaches in the bay which could hold five vehicles compared with seven coaches alongside the main platform. *Lens of Sutton*

Below: 'West Country' class 4-6-2 No 34104 *Bere Alston* shunts the empty stock from a Plymouth excursion at Sidmouth on 3 August 1959. Class 3MT 2-6-2T No 82017, which had assisted on the climb from Tipton St John's, can be seen in the background. *S.C. Nash*

Above: Ivatt 2-6-2T No 41292 has steam up in readiness to depart from Sidmouth with the 5.45pm to Sidmouth Junction on 7 July 1962. *H. C. Casserley*

at Exmouth Junction. The turntable, now seeing little use, was removed for scrap in the 1920s and, with the shed being seldom used, it was closed in the mid-1930s. This was the only sub shed of Exmouth Junction to be closed under the jurisdiction of the Southern Railway. The tracks leading to the shed were cut short and a stop block installed; a supply of emergency coal was stocked if needed, and locomotives still used the former shed road to take water.

Locomotives used on the line in the early days consisted of the 'Ilfracombe Goods' 0-6-0s which had been used on ballast trains during the construction of the line. No 283 was involved in an accident at Tipton St John's on 11 May 1876 when it killed a bullock that had wandered on to the track, and the same locomotive was still allocated to the shed in March 1878. The Beattie Standard 2-4-0 well tanks also appeared in the early years, with Adams 'O2' tanks replacing the 'Ilfracombe Goods' on the Sidmouth line by 1890, No 227 being allocated at Sidmouth; Adams 'T1' 0-4-4 tanks, including No 11, also worked on the line, and from 1901, 0-6-0 tanks of the 'G6' class were utilised on the Sidmouth and Budleigh Salterton goods trains, including Nos 270, 274, 275, 278 and 279. One of the ex-LBSCR 'Terrier' tanks purchased for the Axminster-Lyme Regis line, No 734, was in use on the Sidmouth line

in the spring of 1903, and also on the August Bank Holiday of the same year. With the opening of the line to Lyme Regis, one of the spare 'Terriers' often worked the Sidmouth goods. No 734 was allocated to Sidmouth for a short while by mid-1909, when the 'O2s' took over the Lyme Regis services, with sister locomotive No 735 working goods trains from Sidmouth Junction to Sidmouth and Exmouth. A double-framed Beyer goods No 0287, single-framed Beyer goods No 303, Drummond 'K10' No 152, Adams 'Steamroller' No 0383 and Adams Radial No 0107 are all believed to have worked on the line on various specials etc. The previously mentioned excursion trains changed engines at Sidmouth Junction before proceeding to Sidmouth. The forward engines were usually provided by Exmouth Junction, but on some occasions a Salisbury-based locomotive worked through to Sidmouth. Three Adams '0415' class 4-4-2 Radial tanks were allocated to Exmouth Junction in 1913 for the Lyme Regis services, with one of the spare locomotives being utilised to work on the Sidmouth branch at infrequent intervals, while many of the Adams 'O2' locomotives allocated to Exmouth Junction could be found working the various services 'twixt Exeter, Exmouth, Sidmouth and Sidmouth Junction up until mid-1932 when the Drummond 'M7' 0-4-4 tanks appeared. Most of the 'O2s' were transferred to other duties and a number

Above: A view taken from the roadside of Newton Poppleford on 7 July 1959. The station opened on 1 June 1899, two years after the opening of the line to Salterton. *H. C. Casserley*

were sent to Eastleigh for scrap, but a few were still to be found on the line during World War 2, including Nos 193, 224, 230 and 232. Drummond 'L11' 4-4-0s (known as the 'Large Hoppers') and members of the 'K10' class ('Small Hoppers') also worked passenger trains on the branch.

Two 0-6-0 tender locomotives of the '0395' class, Nos 3029 and 3433, had no weekday duties at Exmouth Junction in July 1939 and were utilised on the 8.52am Saturdays only Exmouth Junction-Whimple freight, thence light to Sidmouth Junction to work the 1.18pm to Sidmouth (a through train from Waterloo), returning with the 3.45pm freight to Sidmouth Junction and Exmouth Junction. Designed by William Adams, the two locomotives were once part of a numerous class (70 in total), of which 50 were sold to the Government and sent to the Middle East during World War 1 and never returned. Eighteen survived into BR days, with the last loco being withdrawn in 1959. The restriction on Light Pacifics (except for naming purposes) was removed in 1951 and a speed restriction of 40mph was applied; Nos 34104 *Bere Alston* and No 34011 *Tavistock* were seen on the line in 1959 working passenger and freight trains respectively. 'N' class 2-6-0 No 1840 and 'U' class 2-6-0 No 1807 were at work on the line in 1948.

From the early 1950s the BR Standard Class 3MT and ex-LMS Ivatt 2-6-2 tanks generally monopolised

the services on the branch, although 'M7s' were still to be seen until the late 1950s and early 1960s. DMUs appeared on the branch from 4 November 1963, and larger members of the 'paraffin burners' such as the North British Type 2, and Beyer Peacock 'Hymeks' worked the summer Saturday services to London, although some of the Saturday services in 1964 were still worked by steam locomotives as was the Sunday 9.23am excursion from Waterloo.

Newton Poppleford

The station, situated just east of the village of the same name, opened on 1 June 1899, two years after the opening of the line. Built of brick, the single-storey station building was positioned on the 184ft-long single platform located on the down side of the line, with pedestrian access provided by a barrow crossing at the north end of the station. Staff once included a stationmaster and a porter. Subsequently the staff was reduced to a stationmaster, and later still to a leading porter under the jurisdiction of the Ottery St Mary stationmaster, and then his colleague at Budleigh Salterton. A solitary goods siding operated by a two-lever ground frame served a feed store, and two camping coaches were based here. When the line to Exmouth opened in 1903, Beyer Peacock double-

Above: The station at Newton Poppleford, looking towards Tipton St John's, as viewed from the road bridge. The porters' barrows await what little custom may appear on the next train. Freight facilities were withdrawn on 27 January 1964 and the siding was lifted in the following year. *Lens of Sutton*

Below: Newton Poppleford station as seen from the goods yard on 7 July 1959. Box vans stand on the solitary siding and one can almost smell the aroma of tar oozing from the sleepers in the hot summer sunshine and hear the rustling of the trees in a gentle breeze. The silence between the arrival of trains was broken only by the hum of the occasional road vehicle crossing the railway by the bridge in the background. *H. C. Casserley*

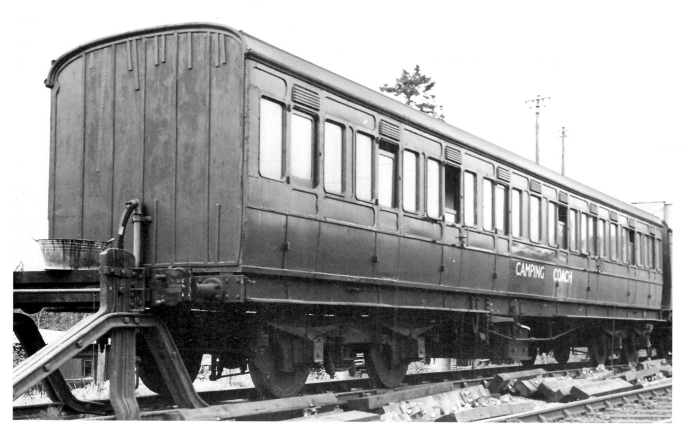

Above: Camping coaches were also stabled at Newton Poppleford as seen here with No S37S, an ex-LSWR 56ft third, on 20 September 1961. Although facilities were sparse in these coaches it must have been nice to wake up in the morning, smell that wonderful fresh Devon air, and perhaps cycle or walk to the nearest farm to buy fresh eggs and milk for breakfast. *A. E. West*

Below: A closer view of the brick-built station building at Newton Poppleford in September 1961. The station originally had a stationmaster and a porter but this dwindled to a leading porter before becoming unstaffed from 16 August 1965 and closing altogether on 6 March 1967. *A. E. West*

Above: Adams 'O2' class 0-4-4T No 199 arrives at East Budleigh with the 11.50am Exmouth-Sidmouth hauling a two-coach set, plus the Waterloo coach on 16 June 1949. *S. C. Nash*

framed 0-6-0s worked the 8.10am goods from Exmouth Junction, arriving at Exmouth at 10.20am and departing for Tipton St John's at 10.45am picking up and setting down wagons at intermediate stations, departing from Tipton at 1.7pm for Exmouth and Exeter. The Summer 1932 freight timetable shows the 10.50am Exmouth to Sidmouth Junction freight shunting here from 1.2pm to 1.17pm. Freight facilities were withdrawn on 27 January 1964, the siding was removed in the following year and the station became unstaffed from 16 August 1965, closing altogether on 6 March 1967. The station site has been completely obliterated.

Colaton Raleigh Siding

The siding, located three miles from and serving the village of the same name, opened with the line in 1897 and was positioned on the up side of the line. The ground frame for the siding was opened by a key on the end of the single-line tablet. East Budleigh was the station in charge of the siding, and the Appendices to the Working Timetable required that a man from that station was to be in charge of the working. Shunting

was accomplished by up goods services; the 10.50am (1934) Exmouth-Sidmouth Junction freight was responsible for shunting the siding from 12.52pm to 12.57pm. Wagons of inward traffic for local farms and estates were delivered by road transport owned by Messrs Miller & Lilley. The siding was taken out of use on 1 February 1953.

East Budleigh

This attractive village of cream-walled thatched cottages is the birthplace of Sir Walter Raleigh (1552-1618) who was born in the thatched manor house of Hayes Barton and the church of All Saints still contains the Raleigh family pew. André Lenôtre, who designed the gardens at the Palace of Versailles, also planned the gardens at Bicton located one mile north of the village. Budleigh station opened with the line in 1897, receiving the prefix 'East' on 27 April 1898. The station, actually nearer the village of Otterton than East Budleigh, was equipped with a single platform 297ft in length on the up side of the single line. A brick-built single-storey building complete with a large awning adorned the platform which also

Above: Class 3MT No 82010 lifts her safety valves while calling at East Budleigh with a Tipton St John's to Exmouth service. *Lens of Sutton*

Below: Two camping coaches were stabled at East Budleigh at the south end of the station. No S34S, an ex-LSWR corridor composite, is viewed on 20 September 1961. *A. E. West*

Above: Ivatt 2-6-2T No 41309 rumbles under the road bridge at East Budleigh with the 1.52pm Sidmouth Junction to Exmouth through coaches on 8 August 1960. *Terry Gough*

Below: The brick-built station building at East Budleigh, with its attractive canopy, was located on a single platform 297ft in length. The station was named Budleigh at the opening of the line and received the prefix 'East' on 27 April 1898. *Lens of Sutton*

Above: Also at East Budleigh on 20 August 1961 was camping coach No S35S, an ex-LSWR 56ft composite. *A. E. West*

Below: No 41308 hauls the 1.28pm Exmouth to Tipton St John's through the lush East Devon countryside at East Budleigh on 9 August 1960. *Terry Gough*

Above: Budleigh Salterton, looking towards Littleham on 8 September 1961. When opened, the station was known as Salterton until renamed Budleigh Salterton on 27 April 1898. The station was the terminus of the line from Tipton St John's until the opening of the extension to Exmouth on 1 June 1903. The up platform to the right, together with a small waiting shelter and the footbridge, were added to the station layout for the opening to Exmouth. *A. E. West*

contained a small brick-built goods shed. A small goods loop with two short spurs and a cattle loading dock were provided at the south end of the station. Access to the loop was operated by a ground frame unlocked by the key on the single-line token. Camping coaches were also stabled here, standing on an isolated length of track to avoid the risk of being bumped when shunting operations were taking place. When the coaches returned to Eastleigh for maintenance in the winter, a short length of track was slewed over by the civil engineers in order to release the vehicles. The station, which became unstaffed on 25 April 1966, was the setting down point for passengers visiting the local beach at Ladram Bay and also the nearby Bicton Gardens with its 18-inch gauge Bicton Woodland Railway. Today, the main station buildings have been carefully restored as a private residence; the platform is complete and the trackbed has been made into a lawn.

Budleigh Salterton

Budleigh Salterton is a small, quiet residential and fashionable resort which has much in common with Sidmouth, with its red cliffs, a pebble beach and a superb sea view over Lyme Bay from West Down Beacon. The sea wall at Budleigh Salterton was the setting for the painting *The Boyhood of Raleigh* by the

Victorian artist Sir John Millais. Opened as Salterton on 15 May 1897, the station here was the terminus of the line from Tipton St John's and was renamed Budleigh Salterton on 27 April 1898. The official opening of the line took place on 14 May 1897. A 'T1' class 0-4-4 tank locomotive, No 359, with Mrs Hugh Williams Drummond (daughter of the Hon Mark Rolle and wife of Brigadier-General Sir Hugh Drummond, Bart, MVO, Chairman of the Budleigh Salterton Railway) on the footplate, arrived at Salterton station with the inaugural train conveying the LSWR officials from Waterloo and other civic dignitaries. The locomotive was adorned with flags, evergreens and a portrait of Mrs Drummond, and was greeted at Salterton by the officials of the town. Sidmouth Volunteers' band played the old favourite 'See the Conquering Hero Comes'. The good lady stepped down from the footplate and 'declared the line open' with her husband giving the formal handover to Mr Wyndham S. Portal, the Chairman of the LSWR. The day was declared a public holiday, the streets and houses suitably decorated, and a sumptuous free tea was provided for all children under 14 years of age. Public services started the following day.

The station originally consisted of a single platform upon which stood a brick-built station building. The layout was fairly basic, consisting of a goods shed, weighbridge and cattle pens. A small wooden engine

The mid-afternoon train from Exmouth, hauled by Class 3MT No 82025, arrives at Budleigh Salterton on 9 July 1959. This particular service terminated here, returning to Exmouth at 3.20pm. The goods shed and yard can be seen in the background.

R. C. Riley

Above: Budleigh Salterton signalbox, seen here on 8 September 1961, was situated at the end of the 316ft-long down platform. The box opened on 1 June 1903, the same day as the extension to Exmouth. *A. E. West*

Left: The up starter at Budleigh Salterton in September 1961 still consisted of a fine LSWR lattice-style post with a lower quadrant arm. *A. E. West*

shed for the branch locomotive was erected at the west end of the yard. Until 1903 the branch was worked on the 'one engine in steam' principle. Services between here and Tipton St John's at the opening consisted of eight trains each way, with one being mixed, plus a separate goods working. The opening of the extension to Exmouth on 1 June 1903 transformed the station from a terminus to a through station, a second platform with a small wooden waiting shelter being added, the original platform then becoming the down platform. Trains from Tipton to Exmouth were designated as 'down'. A ground level signalbox, standing off the end of the down platform, was also opened on 1 June 1903 and controlled the signals, passing loop and entry to the sizeable goods yard which was located on the down side of the layout at the Littleham end of the station. A new footbridge was erected to serve both platforms.

Train services in 1909 comprised 10 trains between Tipton and Exmouth, plus a working from Tipton to Budleigh and a return working between Exmouth and Budleigh Salterton. In the up direction there were 10 trains, three of which had an extended

Above: BR Standard Class 3MT No 82018 trundles along bunker first with the 6.39pm Tipton St John's to Exmouth near Budleigh Salterton on 9 August 1960. *Terry Gough*

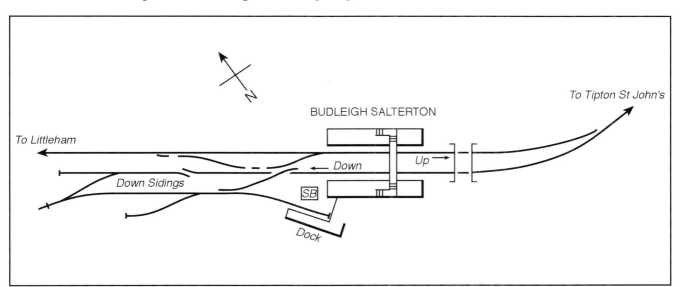

wait at Budleigh Salterton, while three return services ran between Exmouth and Budleigh Salterton on Sundays. The staff at one time consisted of a stationmaster, signalman, clerk, checker, two porters and a porter signalman. Staff numbers were reduced in later years, and in 1966 there was a signalman, clerk and a relief man. Coal traffic for the gasworks was dealt with and, at one time after World War 1, as many as four vans of herrings would be sent to London tacked on to a passenger train. A speed limit of 40mph existed between Budleigh and Exmouth, increasing to 50mph onwards to Tipton St John's except for a 40mph limit through Newton Poppleford.

Stone traffic from Black Hill Stone Quarry arrived by motor transport and was loaded into wagons destined for Portishead. At one time trains were routed to Exeter via Exmouth, but it was found easier to reroute them to Exeter via Sidmouth Junction, usually hauled by a Class 3MT 2-6-2 tank or a 2-6-0 Mogul. 'West Country' class No 21C114 *Budleigh Salterton* received her nameplates in a ceremony held at the station on 26 June 1946. The locomotive arrived via Tipton St John's, thereby avoiding the restriction on the viaduct over the River Clyst on the Exeter-Exmouth branch. The station closed to goods traffic on 27 January 1964, the sidings being abolished on

Above: No 82013 approaches the Budleigh Salterton distant signal with the 5.45pm from Exeter, also on 9 August 1960. *Terry Gough*

2 February 1965, but the signalbox lasted until 6 March 1967. The station site has been developed into a supermarket and a housing development.

Budleigh Salterton Locomotive Shed

The wooden locomotive shed, measuring 50ft x 18ft, was of a simple construction with an inspection pit located inside the building and a wooden water tank and coal stage situated outside. One of the 2-4-0 Beattie standard well tanks, No 253, worked one of the earliest trains to Budleigh Salterton and was outstationed here in 1890. The shed was responsible for all the Budleigh Salterton-Sidmouth Junction workings, as the Board of Trade insisted that 'only tank engines are to be used'. The locomotive was returned to Exmouth Junction weekly for repairs and boiler washout. The extension to Exmouth, from June 1903, inevitably led to the closure of the shed, although it was retained for a time to service the locomotives working the early and late services starting and finishing at Budleigh, until timetable

alterations put an end to this practice with traffic becoming the responsibility of locomotives from Exmouth shed. The shed, which was out of use by 1911, was left standing until it was demolished in 1925, although the engine sidings remained in situ until closure of the line in 1967.

Littleham

The station, constructed on a sharp curve, was sited approximately half a mile from the village and lay on the outskirts of Exmouth, opening with the new link from Budleigh Salterton on 1 June 1903. Having platforms of 402ft in length, the main brick-built station buildings, complete with a substantial canopy, adorned the up platform, while the down platform contained a small timber-built waiting shelter. The station was equipped with a passing loop, signalbox and crossing gates. Four sidings on the down side featured a cattle dock, coal staithes for the local merchants and a goods shed. A shunting neck was also provided. The signalbox was positioned at the Exmouth end of the down platform, and had a knee-high frame, plus a gate wheel for the level crossing.

Above: Littleham station, along with the extension to Exmouth from Budleigh Salterton, was opened on Whit Monday, 1 June 1903. This view, taken on 26 September 1966, shows the main station building situated on the up platform with the level crossing gates open to road traffic. *A. E. West*

Below: Littleham station, viewed from the level crossing on 10 July 1962. Part of the signalbox can be seen to the right and the up platform is to the left in this scene looking towards Tipton St John's. Much holiday traffic for the nearby Sandy Bay caravan camp was dealt with here. *H. C. Casserley*

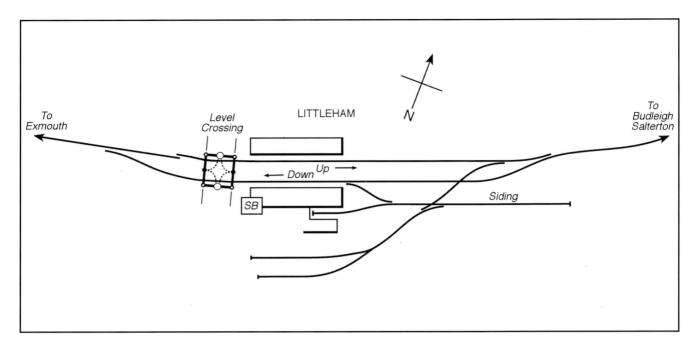

Passenger traffic for the nearby Sandy Bay holiday camp was quite substantial at times, and there was regular business from local inhabitants travelling to Exmouth and Exeter. Staff in the early years consisted of a stationmaster, clerk, signalman, goods checker and a porter. As with most branch stations, the signalman was called upon for station duties when he was not required in the signalbox. From the 1920s, rationalisation reduced the staff to two signalmen and a porter who came under the jurisdiction of the Budleigh Salterton stationmaster. In later years, staffing consisted solely of a signalman who was also responsible for issuing tickets, for which purpose a wooden extension was added to the signalbox in 1961. Camping coaches were also kept at the station in the summer months — varying from two to four at any one time, until 1964. The stock off the Saturdays only Cleethorpes-Exmouth service was stabled at Littleham throughout the week. Track alterations occurred on 25 June 1940 when the No 7 loop points at the Budleigh end of the layout were moved 70ft nearer to the signalbox, with No 11 points leading from the up loop to the goods yard removed at the same time.

In 1951 there were 10 trains each way on weekdays, 11 each way on Saturdays, and four trains in each direction on Sundays, all services travelling to, through or from Sidmouth. Sunday services were withdrawn in the winter, this remaining the situation until closure of the line. With the arrival of diesel services in 1963, the Monday-Friday through coaches were withdrawn, the 1965 summer timetable witnessing the final through services on the Tipton-Exmouth line. Goods traffic ceased on 27 January 1964 and the signalbox closed on 8 May 1967. The only remaining feature at Littleham is the stationmaster's house standing amidst houses and bungalows. The lines between Sidmouth Junction and Sidmouth and Tipton St John's and Exmouth closed on 6 March 1967, leaving the Otter Valley

bereft of the best public transportation system that it would ever have. Track lifting started from the Sidmouth Junction end on 28 May 1968, the branch having been severed at the junction with the singling of the main line on 11 June 1967. The demolition train gained access via Exmouth and Tipton St John's and recovered materials were returned to Exeter via Exmouth. The River Otter burst its banks on 10 July, playing havoc with the track lifting schedule by breaching the line at Newton Poppleford and East Budleigh. By this time, the track had been lifted as far as Ottery St Mary, and work then progressed from East Budleigh to Exmouth, the work being completed by September. The sections of track affected by the flooding, along with approximately 30 wagons that had been marooned at Tipton St John's, were recovered by road transport.

Footplate Memories

Footplate work was often dangerous and at most times was a hard graft, but the comradeship was, above all, the most important element. And the pay wasn't too bad either, especially with the summer workings when we started rest-day working, which leads me to the following story. A pair of footplatemen from my own shed at Yeovil Town were rostered one day to work a special van train from Exeter with relief at Yeovil Junction. The job involved riding down 'on the cushions' to Exeter, preparing an engine at Exmouth Junction and working home. The normal way of things, upon arrival at Exeter Central (and I have done it many times myself) was either to get a lift to Exmouth Junction on a light engine that was travelling there or catch an Exmouth train, get out at Polsloe Bridge Halt and walk to the shed, entering through the back entrance. When our intrepid duo arrived at Exeter

Above: The level crossing at Littleham as seen from the down platform and looking in the Exmouth direction. Part of the signalbox can be seen to the left with its unusual 'ticket office' sign. *A. E. West*

Below: Littleham signalbox was situated at the Exmouth end of the down platform and opened with the line in 1903. The box was unusual in having a ticket office which was installed in May 1961. At that time, only a signalman staffed the station and this extension to the box saved him trekking over to the old booking office on the up platform to issue tickets. The box closed on 8 May 1967. *A. E. West*

Above: Both platforms at Littleham were 402ft in length, the down having a small timber-built waiting shelter. The roof of the goods shed can be seen in the background. *A. E. West*

Central, as luck would have it, there was no engine at that particular moment travelling back to the junction, but there was a train in the up bay, complete with a Standard tank, having all the hallmarks of an Exmouth-bound working. My two mates climbed aboard, lit their fags, and the train departed. All well and good until they reached Exmouth Junction when the train, instead of turning right on to the Exmouth line, went straight on! As it tore past Exmouth Junction on the main line, our two mates looked at one another in amazement. The penny now dropped; they were on one of the Exmouth trains all right, but this particular one was travelling via Sidmouth Junction! What a carry on, especially if the Exmouth Junction foreman found out that the Yeovil crew were absent. Eventually the brakes started to grind on the coach wheels as they arrived at Pinhoe and the signalman looking out from his box window by the level crossing was quite surprised to see two footplatemen climb out of the train and walk towards him. They were becoming worried — how were they going to get back to Exmouth Junction in time to prepare the locomotive? Only an hour was allowed to prepare and move the engine off shed, and if they had to catch a train back to the Central station, they would still have to travel to Exmouth Junction.

The signalman shouted down from his window, 'What do you boys want?' in his broad Devon accent. Sheepishly the duo entered the box and explained to him what had happened. He couldn't help laughing, 'well, there is no down train for about an hour' he explained. However, the faces of the two men lit up when he mentioned that a light engine en route for Exmouth Junction was on its way from Sidmouth Junction, and at that moment the bell rang from Broad Clyst asking for the road. Without more ado, the two footplatemen wandered over to the down platform, and not long after, the light engine hove into sight, the crew wondering why they had been admitted under the 'yellow'. They stopped at the platform and picked up the errant crew and set off for Exmouth Junction. Upon arrival, my two mates wandered to the booking-on window, the clerk giving them the number of their engine and mumbling something about Yeovil men being late. And that was that! They set to work on the engine as fast as they could and, with only minutes to spare, reversed into Exmouth Junction yard at the allotted time with the blower hard on to burn through the fire. Their journey back to Yeovil Junction went according to plan without incident. Unfortunately for them, the story soon spread around the signalboxes and, of course, the news had reached Yeovil Town before they did!

Above: The 8.50am Tipton St John's to Exmouth, hauled by No 82018, approaches Knowle Bridge near Littleham on 12 August 1960. *Terry Gough*

Above: Classic 'South Western' elegance at Exeter Queen Street in 1864. 'Improved Clyde' class 2-4-0 No 75 *Fireking* entered traffic from Nine Elms Works in July 1864 and is pictured here in original condition with 7ft driving wheels, the final pattern coal-burning firebox, concentric tubular feedwater heating, 16½in by 22in cylinders and a 1,950 gallon six-wheel tender. This locomotive, in company with sister locomotives *Fireball* and *Firebrand,* went new to Nine Elms shed and was used on the Southampton expresses. It also appeared regularly on the Waterloo to Salisbury services and occasionally on the Salisbury to Exeter trains. Allocated to Northam in March 1878, *Fireking* was finally withdrawn from Dorchester shed in May 1888 having attained a total of 769,575 miles. Locomotive footplatemen are a hardy breed and none more so than the early LSWR crews. I know personally what it was like to work on a footplate climbing Honiton bank during severe weather — but conditions must have been dreadful on an exposed footplate such as this. *Author's collection*

Below: Wagons of various railway companies including the Midland Railway line the goods yard at Exeter Queen Street, with the gaunt buildings of Exeter Prison dominating the skyline. *Lens of Sutton*

Exeter Central-Exmouth

A meeting had been held at Exeter on 19 January 1825 to consider 'the propriety of laying a rail-road for the more expeditious and economical conveyance of goods from the mouth of the Port of Exeter.' Events had been brought to a head by businessmen dissatisfied with the fact that larger ships could not use the Exeter Canal, which had been in use since its opening in 1566. Other meetings were convened, with the cost of construction of a line to Exmouth being estimated to be £50,000. Not yet determined, though, was the intended course of the line. One proposal was for a route on the western bank of the River Exe via Exminster and Starcross, which it was claimed would bring in only £1,000 annual income, while the other was for a line on the eastern bank through the towns of Topsham and Lympstone which would generate more traffic, resulting in revenue of £3,000 per annum. It was decided, but not with everybody's consent, to use the eastern bank. Much wrangling and discussion ensued, but the scheme was abandoned when Exeter Corporation announced its intention to extend the canal to Turf, thus enabling the canal to be used by vessels with a 14ft draught and 400 tons in weight. The cost of the new works totalled £113,355 and work started on 20 April 1825.

Railway mania began in the area on 4 August 1845 when a scheme by the Exeter, Topsham & Exmouth Railway was announced, followed three days later by another company issuing a prospectus. Both companies bore the same name and, to save confusion, in true railway fashion, they were known by the respective names of their company secretaries, 'Dawes' and 'Heads'. By contrast, the Great Western Railway entered the fray at the end of the year with its prospectus for the Great Western & Exeter, Topsham & Exmouth Junction Railway, running from St Davids station in Exeter. Another powerful foe, in the shape of the South Devon Railway, proposed to join with the 'Heads' company to construct a railway to Exmouth, running over the SDR line to the atmospheric pump house north of Exminster (then under construction), crossing the Exeter Canal and spanning the River Exe on a 14-span viaduct to Topsham and thence to Exmouth. The GWR scheme had come to nothing by 1846, leaving it to support its broad gauge ally, the South Devon. The proposed Exmouth company proposals were now virtually running neck and neck with support from all sides. The 'Heads' company now had support from none other than Brunel himself, advocating the adoption of the atmospheric system such as was used on the South Devon. All to no avail, however, as the narrow gauge proposal of the 'Dawes' faction won the day. The intended line was to run along the eastern bank of the Exe from Exeter city gaol to a terminus at Exmouth, with a branch running to Parker's Quay at Topsham. Joseph Locke was appointed the engineer, with Thomas Whitaker, surveyor of bridges to the County of Devon and surveyor to the City of Exeter, as the resident engineer. Authorisation was given by Parliament on 3 July 1846 under the title of the Exeter & Exmouth Railway, which had powers to sell or lease the line to the LSWR's Exeter, Yeovil & Dorchester Railway, although that particular scheme was itself to be abandoned due to shortage of money. The Exeter & Exmouth was now left high and dry as its whole proposal depended on leasing its route to the LSWR.

Exmouth, by the late 1840s, had a population of approximately 5,000. The South Devon Railway had opened to Teignmouth on 30 May 1846 and onwards to Newton Abbot on 30 December the same year. Broad gauge trains could be seen on the other side of the Exe, and the inhabitants of Exmouth wanted a rail communication, all the more so as trade in the town was declining. Two more attempts were made in 1851 to provide Exmouth with a rail link. The broad gauge camp reappeared with a further proposal to link Exminster with Exmouth, while the 'Dawes' scheme to link Exmouth with the LSWR was revived, the LSWR having successfully gained Royal Assent for its Central Route (the Salisbury & Yeovil Railway) and the Coastal Route (the nominally independent Exeter, Yeovil & Dorchester) on 22 July 1848. However, there were deep divisions within the LSWR itself, as some of its shareholders wanted the central, some the coastal route, and others wanted neither! At a meeting held by the South Western directors on 26 October 1852, the voting was 12,610 to 12,389 against the Salisbury & Yeovil. Meanwhile, back at Exeter, in 1852 an alternative broad gauge line had been proposed along the east bank of the Exe. The line was to run from The Point, Exmouth, to Lympstone on an embankment, the idea being to

Above: Exeter Queen Street, showing the improvements made in 1925 when the up platform was extended to a length of 1,210ft, thus enabling it to accommodate two trains simultaneously on the main line face and creating an up bay on the opposite face. It was not until the second stage of reconstruction, when a total rebuild of the station was completed, that the station was formally opened and renamed Exeter Central on 1 July 1933. *Lens of Sutton*

reclaim hundreds of acres of land from the sea, thereby increasing profitability for the shareholders. Despite this imaginative scheme being put forward again in the following year, it did not succeed.

The scheme for the Exeter & Exmouth appeared in December 1853, endeavouring to raise the sum of £160,000 in £20 shares for the 'narrow gauge' route. The citizens were very interested in this proposal, especially when it was revealed that no more than 10s per share would be used on Parliamentary expenses. By January 1854, of the £20,000 that Exmouth had indicated it could raise, the sum of £15,000 had been promised. The broad gauge camp, however, with the support of Brunel and Robert Wreford (an Exeter solicitor) and many other supporters, was still a dangerous adversary, for in the same year a proposal, similar to that of 1846 for crossing the Exe and the canal from the South Devon Railway broad gauge line at Exminster reappeared. A new addition to the former proposal was for a separate route from the SDR to Exeter quay. The townspeople of Exmouth, believing the LSWR had forsaken them, were now impressed sufficiently by the South Devon's proposal to place

their allegiance with the broad gauge. The LSWR, by contrast, had been dragging its heels, and despite supporting the Salisbury & Yeovil, the Act receiving Royal Assent on 7 August 1854, had achieved nothing further. This resulted in the Act of 2 July 1855, the Exeter & Exmouth Railway, authorising Brunel's broad gauge line to Exmouth via Exminster and Topsham. The Government at the time wanted better transport facilities between the naval and military establishments of Plymouth and Southern England, and when the LSWR applied to Parliament in 1855 with an application for extra time to complete its proposed Andover-Basingstoke line, it was informed in no uncertain terms to honour its pledge to build a 'narrow gauge' route to Exeter. This resulted in the Parliamentary Committee inserting clauses into the Bill, by which the LSWR was bound, under the penalty of stoppage of their dividends, to introduce and use their best endeavours to pass through Parliament a Bill for a 'narrow gauge' line to Exeter. Early in 1856 it was announced that 'an intimate connection between the South-Western and the Salisbury & Yeovil companies is now established'

198

Above: Allocated to Exmouth Junction shed in June of the same year, 'N15' 4-6-0 No E740 *Merlin* awaits its passengers to board before departing for Waterloo from Exeter Queen Street on 8 August 1928. This locomotive was converted to oil firing on 14 December 1946, reverting to coal burning on 30 October 1948. It was withdrawn from service in December 1955 and broken up at Eastleigh Works. *H. C. Casserley*

and, by the summer of 1856, considerable progress had been made on the Salisbury & Yeovil works. On 21 July the same year, the LSWR obtained its act for the Yeovil to Exeter line; the 'narrow gauge' was, at last, advancing to Exeter.

Meanwhile back at Exmouth, the ceremony of turning the first sod took place at the site of the intended Exmouth station on 27 November 1856, and within a week of the ceremony the route of the line between Exmouth and the River Clyst near Topsham was being marked out by Captain McNair for Brunel himself. On 1 August of the following year, the directors of the Exeter & Exmouth announced that they had arranged with the Bristol & Exeter Railway and the South Devon to lease the line for 10 years at £3,000 per annum, but not including the line to the quay at Exeter. The shareholders now threw the proverbial spanner in the works by appointing a committee, which was intent on reducing the total costs of the project estimated at £94,435 (exclusive of the Exmouth quay branch). The report of the committee rejected the agreement with the B&E and SDR, preferring the course of the LSWR line, and

now ordered the earthworks to be proceeded with between Exmouth and Topsham. The Yeovil & Exeter Railway (LSWR) resolved to construct a branch line between Exeter and Topsham, with the Exeter & Exmouth continuing the line from Topsham to Exmouth. The line, when completed, was to be worked as usual by the LSWR for 50% of the gross receipts, of which the total sum was to be determined by the respective mileages of 5 miles 31 chains to the South Western, and 5 miles 43 chains to the Exeter & Exmouth, who also gained by not having to pay for the bridges crossing the Exeter Canal and the River Exe. Both companies were to share the cost of constructing Topsham station and the line down to the quay. A new Act had to be applied for, with powers to abandon part of the authorised line (the 1855 Act) to make new works and reduce the capital from £70,000 to £50,000, and borrowing powers from £33,000 to £16,000. This was achieved in the Act of 28 June 1858. The LSWR also gained Parliamentary powers to construct a branch to Topsham under the Act of 12 July 1858, and in due course the LSWR amalgamated with the Exeter & Exmouth, the

Above: The buildings of Exeter Prison loom over Exeter Central as 'M7' No 668 performs station pilot duties on 28 August 1945. *H. C. Casserley*

required Act authorising this being granted on 5 July 1865, with amalgamation taking place on 1 January 1856. Thus was obtained one of the most financially viable branches that the LSWR ever owned, becoming the only branch line to have an intensive commuter and holiday service anywhere in the West Country. J. E. Errington was appointed engineer of the line, and W. R. Galbraith as resident engineer. The contractor for the Topsham to Exmouth line was James Taylor of Northernhay, Exeter, and the line was to be built at a cost of £39,000, this figure not including the telegraph and stations. The line was to be ready by 1 May 1860.

The only engineering difficulty on the line was the viaduct across the River Clyst, the cofferdams for which had been started in August 1859. The foundations for the viaduct piers had to be sunk to a depth of 25ft due to the intemperate and shifting nature of the river bed. Some 500 men and 50 horses were employed on the line, but the early months of 1860 brought storms and high spring tides to the West Country, thus delaying deliveries by sea of stone from Babbacombe for the bridges and culverts etc. By August of the same year, two of the five miles of line had been completed, but another severe storm in the same month was responsible for washing away hundreds of tons of ballast. The line was eventually

completed, with Colonel Yolland inspecting the line for the Board of Trade on 27 April 1861. His reports feature the branch in two parts: the Topsham branch of the LSWR, and the Topsham to Exmouth section of the Exeter & Exmouth. Stations were at Topsham, Woodbury Road, Lympstone and Exmouth. The opening day, 1 May 1861, was declared a general holiday in Exmouth, and the first train left Exeter Queen Street at 7.46am, consisting of 11 four-wheeled carriages containing 150 passengers, and hauled by 2-2-2 Beattie well tank No 36 *Comet*. A second train departed from Exeter at 10.45am and was over-subscribed, including the Exeter Artillery Company Band, and so many people were left on the platform when the 17 or 18 carriages departed, that a further train had to be obtained to convey them to Exmouth. It was originally planned that the directors of the line would travel on the 10.45am, but apparently since the welcoming committee at Exmouth would not have been ready at that time, a special train started at 12 noon for the people left behind by the second train. The special conveying the directors of the line, officers of the LSWR and other worthies, arrived at Exmouth to a tumultuous welcome at 1pm. The Exmouth Band preceded various volunteer corps from the station; the town was

Above: Stroudley E1/R 0-6-2T No 2697 (originally LBSCR E1 class 0-6-0T No 697) stands alongside the up platform at Exeter Central after banking a freight up the 1 in 37 gradient from Exeter St Davids on 28 August 1945. *R. M. Casserley*

packed with happy people, with flags and mottoes hanging from every house and shop. Triumphal arches were covered in inscriptions such as 'Our Gallant Volunteers', 'Welcome to Visitors', and 'Increase of Trade' etc and, as usual with such proceedings, the well-to-do were entertained to a banquet in the Globe Hotel. During the first five days on the branch, a total of 2,000 passengers a day used the line, and the number of tickets issued by 31 August 1861 totalled 80,000. LSWR railmotors appeared at Exeter in 1906 with new halts opening at Lions Holt and Mount Pleasant Road for the new services to Honiton. The line was doubled between Exmouth Junction and Topsham two years later, on 31 May, and two new halts opened at Polsloe Bridge and Clyst St Mary & Digby for the new railmotor service between Exeter and Topsham.

Exeter Central

Opened to public traffic as Queen Street on 19 July 1860, the station was a terminus until the opening of the connection to the GWR station at Exeter St Davids and onwards to Crediton, in 1862. It was always described as dark and gloomy under the imposing

overall roof and was subject to constant criticism over many years. Two platforms served the main line trains. Various improvements were made to the track layout, including the lengthening of the up platform, and a new signalbox was added in 1925, but it was the Southern Railway that was ultimately responsible for reconstruction work, which started in 1931, resulting in a superb new station and improved layout. This was formally opened and renamed Exeter Central on 1 July 1933.

Trains on the branch in the early years were in the hands of the Beattie 2-2-2 and 2-4-0 well tanks. *Comet* has already been mentioned as working on the opening day, and a 'Nelson' class 2-4-0 well tank, No 144 *Howe*, recorded as being used at Haslemere for banking duties in March 1860. This locomotive was transferred to Exeter in April 1861 for working goods trains on the branch, but was transferred away from Exeter in 1875. Various 2-2-2 well tanks of the 'Sussex' class, including Nos 4 *Locke,* 19 *Briton* and 20 *Princess*, all three dating from 1852, also worked on the branch. 'Tartar' class 2-2-2 well tanks Nos 2 *Tartar,* 12 *Jupiter,* 13 *Orion* and 33 *Phoenix* were in use between September and December 1867, being joined in 1868 by fellow class members Nos 17 *Queen* and 18 *Albert*, making

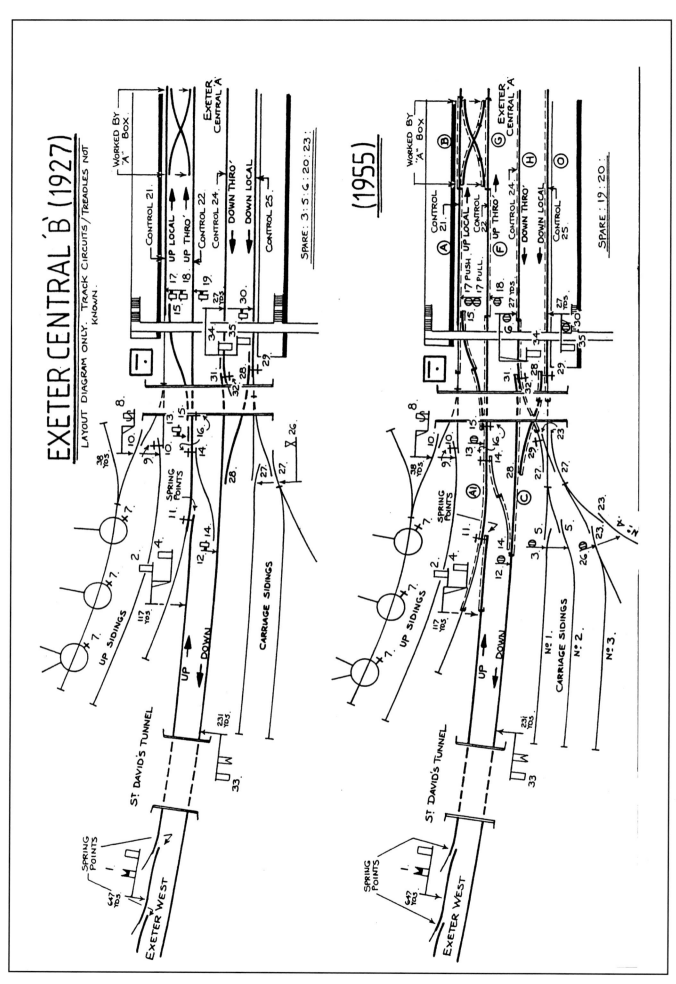

EXETER CENTRAL 'B' (1927)

LAYOUT DIAGRAM ONLY. TRACK CIRCUITS / TREADLES NOT KNOWN.

(1955)

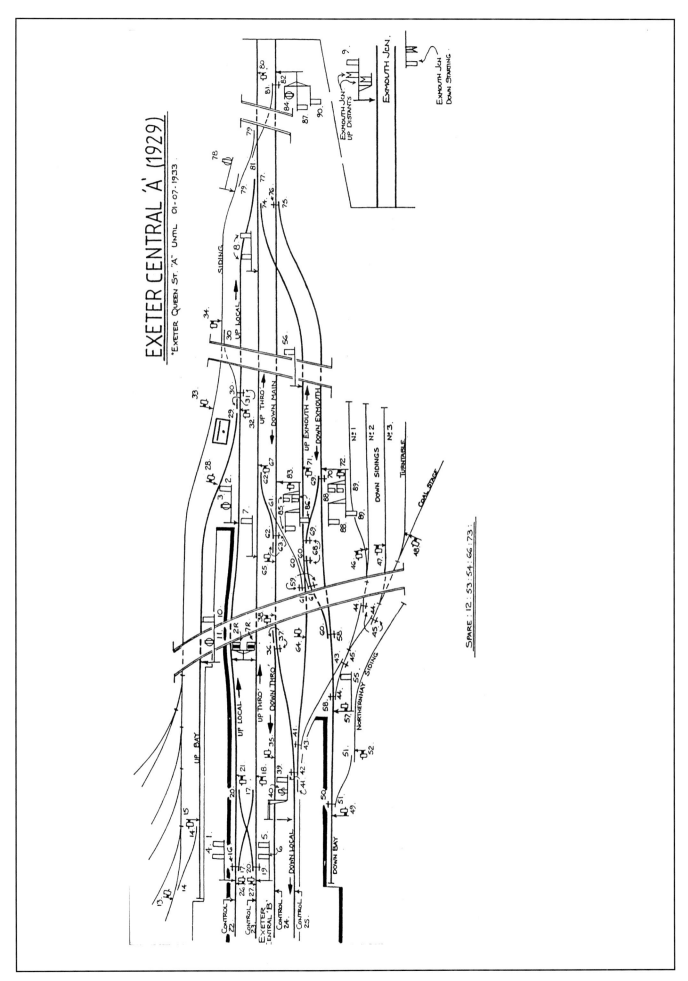

EXETER CENTRAL 'A' (1929)

"Exeter Queen St. "A" Until 01-07-1933

SPARE : 12 : 53 : 54 : 66 : 73 :

203

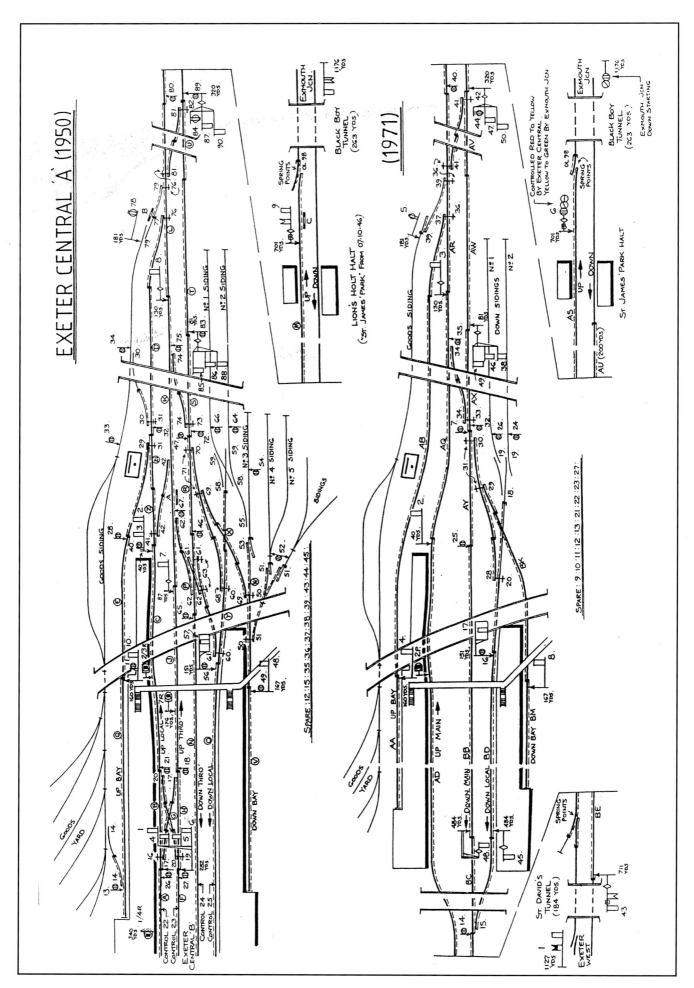

Above: No 82013 stands in the up bay at Exeter Central while the guard fills out his journal with the engine number and driver's name etc for the 2.15pm to Exmouth on 10 September 1958. *Terry Gough*

all six members of this class active in the Exeter area in 1867-8. The Beattie well tanks, although small in stature, performed well on the Exmouth branch, hauling trains of four-wheeled stock out of Queen Street and storming up the 1 in 100 gradient towards Blackboy Tunnel before turning off the main line (doubled in 1870) and striding down to Topsham. Displaced from the main line for various reasons, including age and condition, some of the 1846-8 Rothwell 2-2-2 tender engines came to the branch. Built by Rothwell of Bolton, Nos 73 *Fireball* (the first of the class), 77 *Wildfire* (originally named *Harpy*, until May 1852) and 97 *Pegasus* all appeared. Other tender engines used in the formative years included the Fairbairn 2-2-2 singles (constructed by W. Fairbairn of Manchester from 1846-8) such as Nos 64 *Acheron* and 65 *Achilles*. All members of the class were withdrawn in the late 1860s except the two aforementioned locomotives which lasted at Portsmouth until 1871-2. By 1878, Exeter had a large number of the Standard 2-4-0 well tanks allocated, most being employed at various times on the branch as well as on other duties. Exmouth Junction shed had opened on 3 November 1887, replacing the former locomotive depot at Queen Street, and six of the rebuilt 2-4-0 standard well tanks were there in the mid-1890s for use principally on the branch.

Coaching stock at first comprised a motley collection of four and six-wheelers transferred from other areas of the South Western. Not surprisingly, due to the growing passenger trade on the branch, this led to some protestations, the directors of the LSWR receiving a complaint from the townspeople of Exmouth on 19 January 1899 regarding the 'antiquated and uncomfortable coaches' on the line. Things improved with the disappearance of the older stock, and from the 1930s the rolling stock consisted of nothing more than a miscellaneous assortment of LSWR, LBSCR and SECR stock, including LSWR Warner 56-58ft two-coach lavatory sets and loose thirds. These survived until 1956 when the Swindon-built 63ft non-corridor coaches appeared, thus bringing an inner city suburban feel to the branch. Sets consisted of three, five or seven-coach combinations according to type of service, time of year, rush hour, holiday traffic, etc. The norm during peak-hour traffic would be five coaches; seven coaches was the maximum load permitted during the heavy summer traffic, but three or four coaches sufficed at other times. Maunsell sets also appeared right up until 1964. The services were not the preserve of non-corridor stock, as the 8.10am Exmouth to Exeter Central was formed of two non-corridor seconds and a Bulleid corridor three-set, with the three-set being worked onwards to Plymouth.

Above: 'King Arthur' 4-6-0 No 450 *Sir Kay* stands at the head of an up train at Exeter Central on 28 August 1945.
H. C. Casserley

The inaugural services to Exmouth comprised five trains each way, including Sundays and, from 1 July the same year, trains were increased to seven each way with four on Sundays. The pattern of services was now set to increase from year to year, until by late summer 1887 there were 10 down and 11 up trains, with the Sunday service now having five trains. From 1905 a nonstop service left Exmouth at 9.30am with a booked arrival at Queen Street of 9.30am, the corresponding return service departing from Queen Street at 6.12pm. Two Drummond 'H13' steam railcars, Nos 5 and 6, were allocated to Exmouth Junction, taking over the Exeter-Whimple-Sidmouth Junction-Honiton services on 26 January 1906, and also the shuttle services between Queen Street and Topsham from 1 June 1908 introduced with the doubling of the line. The new railcar services consisted of 10 return daily trains, plus five on Sundays. The railcars, improved and modified from the earlier 'H12' class, must have been an exquisite sight when new, shuffling towards Blackboy Tunnel with their wheels, cylinders and power bogie in green complemented by the coachwork painted in the superb LSWR livery of salmon and dark brown replete with gold lining. The weekday railcar service to Topsham was reduced to nine each way in 1909, with five on Sundays. Railcars Nos 13 and 14 were at Exmouth Junction in 1906 and then despatched to work the Wadebridge-Padstow and Wadebridge-

Bodmin services, while Nos 6 and 14 were recorded on the Exeter-Topsham services in the summer of 1914. All the 'H13s' were withdrawn in 1916 (except Nos 3 and 4 which had gone in July 1919) and the shuttle service to and from Topsham was abolished, although a 'short' service to Topsham augmented the branch trains in 1963.

At the Grouping, the Southern Railway provided a service of 20 down and 21 up trains, increasing to 27 trains in each direction in 1937, plus 19 on Sundays; while in 1948, the summer service comprised 23 down and 24 up trains, with 13 each way on Sundays. Services reached their zenith in 1963 with 31 down trains and 31 up (plus 31 on Saturdays), and with 18 up and 17 down trains on Sundays. Exmouth branch trains arrived and departed from the bay platforms, Nos 1 and 4, at Exeter Central. The Standard and Ivatt 2-6-2 tanks gave sterling service (as did the 'M7s') on the branch trains, barking away from the station with seven coaches of green non-compartment stock during the rush hour, heading swiftly for St James' Park Halt and the tunnel, or gliding into Platform 1 with a morning train, the carriage doors opening to let the commuters, shoppers and schoolchildren troop along the platform towards the exit. DMUs appeared on the branch in public service from 15 July 1963, working as either three or five-car sets. But as usual with the inefficiencies of the management of the day, the timetable was altered in September of the same

Above: Holidaymakers gather their luggage together as Bulleid Pacific 4-6-2 No 21C105 arrives with an up service at Exeter Central on 28 August 1945. A fresh engine will take the train onwards to Waterloo. The locomotive is not carrying nameplates but will eventually be named *Barnstaple*. A Drummond 'M7' working an Exmouth service can be seen in the up bay, with the goods shed beyond. *H. C. Casserley*

year, when three-car sets were introduced, causing congestion on the popular commuter business services. The situation was redressed by an additional trailer being added from 27 January 1964. Main line services to and from Waterloo had been largely handed over to diesel traction in the winter timetable of 1964 and steam power was waning, even more so with the Western Region in charge, leaving the only remaining steam service on the branch, the 3.42pm Exmouth freight, which ceased on 24 May 1965.

The diesel services worked by the first generation DMUs plodded on, until replaced in 1985 by the lightweight Class 142 'Pacer' or 'Skipper' two-car units. These worked on public service trains from 20 January 1986, but problems with the trains operating track circuits and 'disappearing' from the panels in the signalboxes caused them to be withdrawn from 11 March the same year, until returning to full working from 12 May. Still proving troublesome and beset with a number of problems, they were withdrawn during the autumn of 1987, all being sent to Neville Hill depot at Leeds. DMUs of various classes and condition were obtained from depots far and wide and sent to work on the branch, including appearances in 1990 by Class 155 units, although the old 'Heritage' DMUs still soldiered on until 1993. The services today are worked by Class 153 single railcars, interspersed with Class 150/2s. Class 158 units have also worked on the branch,

operated by Wales & West. The weekday services in 1999 comprised 30 down and 29 up trains, with 13 trains each way on Sundays. Some of the services are run as semi-fast by omitting St James' Park Halt, Polsloe Bridge and Exton, accomplishing the journey in approximately 24 minutes, with the trains running on to either Barnstaple, Crediton, Newton Abbot or Paignton. Passenger traffic with commuters, shoppers and holidaymakers is still very healthy, and especially so with the daily rush-hour traffic by road to Exeter becoming more and more of a nightmare. An innovation started in 1997, and continued in the 1999 timetable by Wales & West, was the three return Sunday services from Exmouth to Okehampton. The once busy station at Exeter Central now springs to life only with the arrival and departure of the Exmouth branch trains and the Class 159 Turbos operated by South West Trains on the main line services to and from Waterloo. The long echoing platforms, disused signalboxes and empty weed-grown sidings bear witness to a more illustrious past.

St James' Park Halt

Opened as Lions Holt Halt on 26 January 1906, and renamed on 6 October 1946, St James' Park Halt is situated in the cutting between Exeter Central and Blackboy Tunnel and was opened especially for the

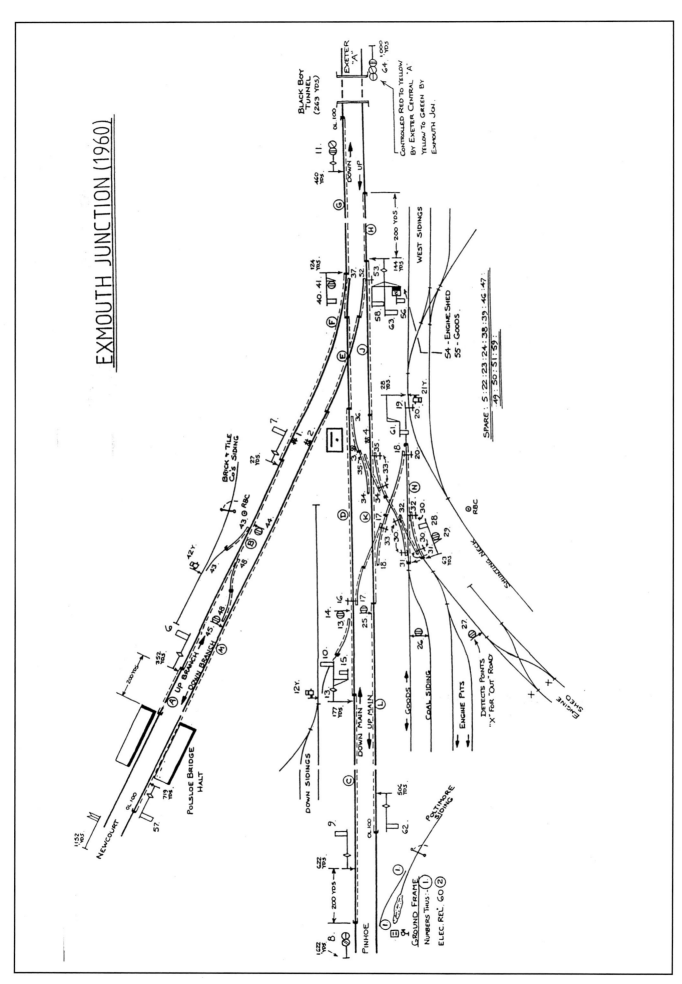

EXMOUTH JUNCTION (1960)

Above: Adams 'O2' class 0-4-4T No E235 approaches Exmouth Junction with the 3.33pm from Sidmouth Junction on 8 April 1928. *H. C. Casserley*

steam railmotor services from Exeter to Whimple and Honiton. From 1908 it was also served by the shuttle services to Topsham. Lying in the shadow of Exeter City Football Club, the station has two platforms, both originally 119ft long, but the down platform was extended to 244ft in May 1928. Today the station is served by the Exmouth branch trains (except for the semi-fast services), comprising 24 weekday trains from Exeter and 23 from Exmouth, plus 13 return trips on Sundays. After emerging from the 220-yard Blackboy Tunnel and breasting the summit of the 1 in 100 from Exeter Central, branch trains swing right from the up main line at Exmouth Junction. A speed limit of 25mph at the junction applies to all branch trains. The signalbox stands in the vee of the junction. The line to Topsham originally single, was converted to a double line on 31 May 1908, but reverted to single track in February 1973. A ladder junction in the modern day practice with a single turnout has now replaced the former double junction. Trains to Exmouth are designated 'down' and 'up' in the reverse direction. A siding located off the up branch once served the Western Counties Brick & Tile Company. Bricks were manufactured from clay extracted from nearby pits and the output of Rougemont Tileries was conveyed from there in large quantities by rail. A crossover installed between the

up and down branch lines was controlled by Exmouth Junction box. The brick siding was taken out of use on 12 May 1967 and Collards Siding, which was a continuation of the brick siding, was closed on 7 January 1963.

Polsloe Bridge Halt

Opened on 31 May 1908 and serving the eastern suburbs of Exeter, this halt, built for the steam railmotor services to Topsham, is situated on an embankment with stairways giving access from the road. The original platforms were constructed of wood, the halt being rebuilt and extended in 1927 using Exmouth Junction-made concrete components, with the original wooden stairways from the public road also replaced by concrete versions. A small concrete-built combined booking office and waiting room was situated on the down platform, with a waiting shelter at the Exeter end of the up platform. The ticket office was eventually closed but it was always a busy station, especially during the morning rush hours with passengers standing three deep on the platform. The station today still sees considerable usage, with the single track now serving the former up platform.

Above: Drummond 'M7' 0-4-4T No 35 approaches Polsloe Bridge Halt with the 4.33pm from Exmouth on 21 May 1935. Thirteen duties were worked in the 1930s by Exmouth Junction 'M7's including Exeter-Exmouth (two engines), Exeter-Exmouth-Sidmouth (five engines), Exeter-Sidmouth-Honiton-Axminster (two engines) (one SO), Exeter-Exmouth-Sidmouth Junction (two engines), Seaton Branch (SO) and Sidmouth-Sidmouth Junction (SO). Dating from April 1898, No 35 was at first allocated to Nine Elms before transferring to Exmouth Junction by 1931 and was at Plymouth by mid-1937 working the Tavistock and Brent local services. Withdrawal from service came in February 1963. *H. C. Casserley*

Clyst St Mary & Digby Halt/Digby & Sowton/Digby Sidings

Clyst St Mary & Digby Halt, constructed of wooden sleepers, opened on 31 May 1908 in connection with the introduction of the steam railcars and the doubling of the line, and served the village of Clyst St Mary to the east, as well as the nearby Digby Hospital. The halt was in need of reconstruction when closed on 27 September 1948. Digby & Sowton station, situated only eight chains from the site of the original Clyst St Mary station, opened on 23 May 1995 and was funded by Devon County Council and Tesco. The basic platform, complete with waiting shelter, stands on the site of the former down branch line near to the adjacent Sowton industrial estate. The station has a car park and is served by a minibus service which runs to Exeter St Davids station. Digby Sidings, located 36 chains from Clyst St Mary and trailing off the up line, opened in 1884 for the transportation of materials to the new hospital then under construction.

Above: The down platform at Polsloe Bridge Halt as viewed from an Exmouth to Exeter Central train hauled by No 82013 on 2 September 1959. Exmouth Junction signalbox can be seen in the distance.
H. B. Priestley — Author's collection

Access was gained by the key on the end of the single-line token and, with the doubling of the line, a three-lever ground frame was installed, worked by a Sykes lock and plunger from the signalbox at Topsham. The track was partially removed in 1888 upon completion of the hospital, leaving a length of five chains for coal traffic destined for the hospital's boilers etc. The siding was the property of Exeter City Council and was provided with a gate, the key being obtained from Topsham box. The 1932 working timetable shows the 2.40pm Exmouth-Exmouth Junction freight booked to shunt the siding from 5.6pm to 5.18pm. All inward goods were dealt with via Topsham station. The siding was taken out of use on 10 January 1957.

Newcourt Sidings

Opened in October 1943 as a temporary siding for the nearby United States Navy Depot, the American servicemen laid their own trackwork with commendable haste. The civil engineering staff of the

Southern Railway installed a trailing connection to the up line controlled by a ground frame and at the same time constructed a shallow embankment in order to link up with the siding. The depot was served by three sidings. A ground level signalbox, constructed in concrete panels, opened on 23 January 1944 and replaced the ground frame. The new box, known as Newcourt, was equipped with a knee-high frame, an illuminated diagram and a closing switch; a crossover and signalling were controlled by the box which was operated by a porter/signalman who was also responsible for checking goods etc. The sidings as expected were very busy during the war years, especially with the D-Day landings, and eventually came under the jurisdiction of the Ministry of Defence as a stores depot for the Royal Navy. In the 1960s, it was shunted by a daily freight working to Exmouth Junction Diagram No 575. This was a daily turn, except for Fridays when a freight to Yeoford and a return stone train to Exeter were worked. Under dieselisation during the 1970s trip freights were controlled by Exeter Riverside Yard. Apart from the

Above: The single-line tablet is held for collection by the fireman as No 82024 rattles over the level crossing at Topsham with the 12.15pm Exeter Central to Exmouth on 12 August 1960. *Terry Gough*

Below: Ivatt 2MT 2-6-2T No 41306 pulls away from Topsham with the 1.45pm Exeter to Exmouth on 12 August 1960. *Terry Gough*

Above: Class 3MT 2-6-2T No 82010 arrives at Topsham with the 1.25pm from Exeter, also on 12 August 1960. A member of the station staff walks across the track in readiness to uncouple the locomotive which will work the train back to Exeter at 1.46pm. *Terry Gough*

war years the only busy moments at the depot came during the Falklands crisis in 1982. The signalbox was reduced to a ground frame on 5 February 1973 upon the singling of the line, and was abolished in 1986 with the removal of the connection to the sidings.

Topsham

The station, dating from the opening of the line, has the inimitable Tite station buildings on the up platform which at one time were complete with a splendid platform canopy. The down platform was equipped with a wooden waiting shelter which also had a canopy. The signalbox, dating from c1875, was situated on the up platform at the Exeter end adjacent to the level crossing which was operated by a gate wheel in the signalbox. In steam days the branch had a maximum speed limit of 50mph as far as Topsham and 40mph onwards to Exmouth. Trains arriving from Exeter normally had to wait outside the station for a few minutes as the interlocking only permitted trains to enter from one direction at a time. When an up train had entered and stopped at the station, the signals would be lowered to admit the down train over the level crossing and into the station. A large goods shed and yard were situated behind the down

platform, access being gained at the Exmouth end of the station. The branch to the quay on the River Exe was 700 yards in length, curving from the station westwards at 90° and was opened on 23 September 1861. The line was then worked only in daylight, with 'B4' class 0-4-0Ts Nos 88, 91 and 92 being allocated to Exmouth Junction especially for this duty. Adams 'O2' class 0-4-4 tanks were the largest locomotives allowed. Goods trains over the quay branch were limited to eight loaded wagons and a brake van. One of the principal commodities handled during the 1930s was guano from South America, but other goods included coal and timber. Traffic never came up to expectations, although the branch survived until 1957, the track being lifted the following year. It was mainly the increased commuter traffic from this station that was responsible for the doubling of the line to Exmouth Junction in 1908. In its heyday the station had a staff comprising a stationmaster, two clerks, two signalmen, a porter/signalman, goods checker, one leading porter and a porter. As there was no rostered shunter, the porters had to undertake this duty. The 23-lever signalbox worked the double track to Exmouth Junction under Sykes lock and block, with a Tyers No 6 tablet to Lympstone, until 16 September 1962, and to Exmouth until 10 March 1968. The branch to

Above: The 1.46pm Topsham to Exeter Central formed of set No 178, hauled by Class 3MT 2-6-2T No 82010, awaits departure on 12 August 1960. *Terry Gough*

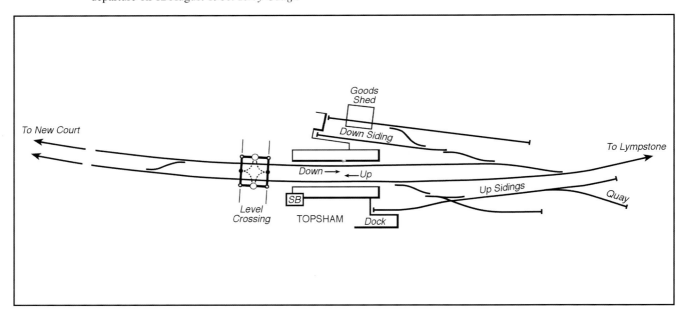

the quay was worked with a wooden train staff until 1909, when it reverted to 'one engine in steam' working.

A Manning Wardle 0-6-0 saddle tank, No 458 *Jumbo*, was involved in a shunting incident at Topsham on 17 January 1896, resulting in the locomotive being withdrawn from Nine Elms five months later with its firebox condemned. On 12 June 1903 ex-LBSCR 'Terrier' 0-6-0T No 734 working an Exeter-Exmouth goods was derailed in the sidings. The block section to the new signalbox at Newcourt,

when opened in 1944, was worked under the three-position open block. The platforms at the station were considerably lengthened in November 1909 and some of the track layout was altered, with trains then gaining access to the quay from the up branch via trailing points to the up siding, instead of the down branch as previously. The sole midday goods service from Exeter to Topsham ceased to run in 1963, the service departing from Exeter Central at 12.55pm with a booked arrival at 1.6pm. The goods yard officially closed on 4 December 1967, with the

Above: The stationmaster and his staff pose for the camera as an Exeter Queen Street to Exmouth train arrives at Woodbury Road in the early 1900s. The station was renamed Exton from 15 September 1958. *Lens of Sutton*

station becoming partially unstaffed from 28 February 1965 and completely unstaffed from 6 May 1968. The goods shed, after being used by a builder's merchant, was demolished in 1993 and the site is now occupied by houses. The line to Exmouth Junction was singled on 5 February 1973, and Topsham became a passing loop on the single line. Lifting barriers came into operation from 20 May 1973, the box closing on 30 January 1988. With this layout, including the lifting barriers, it is now controlled from Exmouth Junction with 'one-train' working to Exmouth.

The station today still has its Tite buildings, although now cement rendered and in private use as offices, segregated from the platform by railings. The signalbox is now a Grade II listed building. It has been superbly restored and looks for all the world as if it is still in signalling use. The line onwards from here to Exmouth was always single track, with trains taking up the Topsham-Lympstone token, following the estuary of the River Exe and crossing the River Clyst viaduct, the spans of which were rebuilt in 1960 using the portion of the original piers that had been built to take double track. This allowed the new trackwork to be built alongside the original without disrupting the traffic, except for a 15mph speed restriction while the building work was in progress. The former trackwork and the remaining girders of the old bridge were dismantled in 1961.

Odams Siding

Odams Siding, serving a fertiliser factory, was situated on the down side of the line immediately beyond the River Clyst Viaduct. Traffic was circulated via Topsham, and the only type of locomotive allowed on the siding was the Adams 'O2'. The key to the gates providing access to the siding was kept in Topsham signalbox; the ground frame was unlocked by a key on the end of the single-line tablet. The 9.18am Exmouth Junction-Exmouth freight was booked to shunt the siding from 11.13pm to 11.18am, and there was also a special trip working from Topsham at 4.8pm, arriving at the siding at 4.11pm and departing at 4.25pm. The locomotive propelled its short train back to Topsham, not exceeding 6mph and with the brake van leading. The train would be brought to a stop before the foot crossing at the Exmouth end of Topsham station and the guard would alight in order to protect any pedestrians who might be using the crossing. The siding was abolished on 25 February 1940.

Exton

The station, serving the village of the same name two miles away, opened as Woodbury Road, remaining thus from 1861 until 15 September 1958 when it

215

Above left: Another view of Woodbury Road in the early 1900s. Passengers gaze at the camera from the carriage windows as visitors walk to the station exit. The buildings are hung with slates to keep out the worst of the winter weather. *Lens of Sutton*

Below left: The 4.25pm Exeter Central to Exmouth leaves Woodbury Road station behind No 82017 on 15 June 1958. *S. C. Nash*

Above: No 82022 heads the 5.10pm Exeter Central to Exmouth near Woodbury Road, again on 15 June 1958. *S. C. Nash*

became Exton. On 28 February 1965 it was renamed Exton Halt, reverting to Exton on 5 May 1969. It consisted of a single platform directly facing the Exe estuary, complete with a two-storey stationmaster's house, with an adjoining single-storey booking hall and waiting room. Two sidings accessed by a ground frame were provided for the local coal merchant. *Firefly*, a 2-2-2 well tank formerly used on the Chard Road-Chard Town branch, was derailed by a faulty point at Woodbury Road on 11 October 1865 and repaired at Exeter Queen Street locomotive shed between March and November 1867. The station was busy during World War 2 when a large Royal Marines training camp was built nearby, and a camping coach was berthed here from 1935 until withdrawal in 1964. Freight traffic ceased on 6 March 1961. Having at one time a stationmaster, the station was staffed by two grade 1 porters from 1918, and became unstaffed from 28 February 1965. Plans were once mooted during the diesel era to provide Exton with an island platform and passing loop, but nothing materialised. The station house is now in private use, with a modern chalet bungalow built alongside. A bus-stop-type shelter is provided for present day passengers, and

trains stop upon request by intending passengers giving a hand signal from the platform.

Lympstone Commando and Lympstone Village stations

With construction starting on 26 January 1976 and opening on 3 May the same year, the single-platform station, complete with the ubiquitous bus-stop waiting shelter, exclusively serves the Lympstone Training Centre of the Royal Marines. Three-car trains can be accommodated at the 224ft-long platform and, as no booking office is provided, a conductor complete with a hand-held ticket machine staffs the station at various times. Approximately 1 mile 3 chains and only two minutes by train after leaving Lympstone Commando, the line passes through a cutting and enters Lympstone, a well-patronised station that opened with the line in 1861

Above: Ivatt 2-6-2T No 41307 departs from Lympstone on the same day in 1958 with the six-coach 12.55pm Exeter Central to Exmouth. *S. C. Nash*

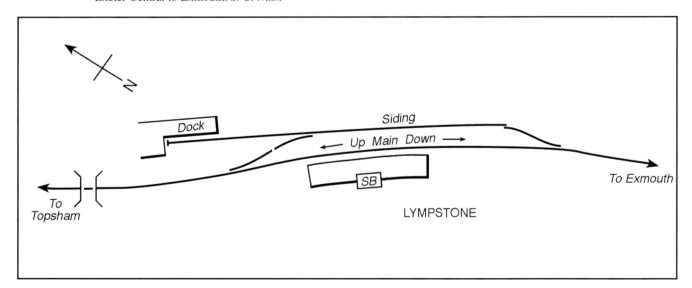

and is located near the centre of the village of the same name. A short, single platform 323ft in length was built, although it could not accommodate a seven-coach train, and a goods loop was also provided as was a short dock siding. The original signalbox stood opposite the platform and was replaced on 29 September 1929 by a new 12-lever box situated in the ticket office. The station building was constructed of brick with a canopy and adjoining this was the timber-built goods and parcels shed. A popular station with commuters and shoppers to Exeter, the station at one time, in the days before the

soulless, unstaffed stations of today, boasted a stationmaster, clerk, two signalmen and a porter/signalman. Subsequently, the staff was reduced to two signalmen who helped with the clerical duties in busy times, before the station became unstaffed on 28 February 1965. Freight facilities, which at one time included coal traffic for the local traders and shellfish which was loaded on to trains up to World War 2, were withdrawn on 4 April 1960. The sidings were removed and the signalbox closed on 16 September 1962. There then followed a period of decline with the station building being

Above: Class 3MT 2-6-2T No 82011, with steam shut off and regulator closed, approaches Lympstone with the 9.18am Exmouth to Manchester on 7 July 1962. *S. C. Nash*

boarded up in 1965 and the parcels shed on the platform serving as a waiting room until all the station buildings were demolished and replaced by a bus-stop shelter. The station was renamed Lympstone Halt on 5 May 1969 and subsequently Lympstone Village on 13 May 1991. Beyond the station, the line traverses the village on a three-arch viaduct then, after a comparatively short stretch of level track, descends for 1½ miles towards Exmouth.

East Devon Brick & Tile Company Siding

Known as the 'Warren' and opened in 1884, the siding serving the East Devon Brick & Tile Company (formally the Western Counties Brick & Tile Co) was situated 1 mile 27 chains from Lympstone. With a facing connection on the down side of the single line it was served by special workings from Exmouth, the locomotive propelling and brake van leading back to Exmouth from the siding when shunting was completed. As with Odams Siding, the only locomotives permitted were the Adams 'O2' tank engines. The ground frame was unlocked by the key on the end of the single-line token. The siding was taken out of use in May 1960.

Exmouth

The original station, as opened on 1 May 1861, was situated in Imperial Road (known later as Station Road) close to the mouth of the River Exe on its eastern bank. Facilities comprised a fairly short island loop platform, part of which had an overall canopy. A run-round loop served the No 1 platform, and goods facilities consisting of a goods shed and three sidings were situated to the west of the station, with a locomotive shed and turntable occupying the east side of the station. A 21.4-chain line from the west end of the goods yard to the site of the intended Exmouth Docks was opened in 1864, the new dock basin opening in 1868. The line from Exeter to Exmouth, when opened, was regulated by 'policemen' who worked the points and signals by adjacent levers. The priority of trains was dictated by the working timetable, amended as necessary by written orders. The electric telegraph was also installed, thus allowing last minute adjustments to be made to the working. Signals were of the revolving disc type designed by Albinus Martin, the disc signals themselves consisting of stop and auxiliary types. Starting signals were not provided, the authority to proceed being given by the policeman, either verbally or by presentation of a train order. Proper signalling as

Above: The terminal building at Exmouth in 1928, four years after its opening on 24 July 1924. This building was constructed in front of the original terminus which was then demolished, enabling a large forecourt to be positioned in front of the new premises. The parcels office is to the left of the main block and the booking office to the right.
Lens of Sutton

we know it today did not appear in the area until c1875, whereupon a signalbox was constructed at Exmouth standing at the throat of the loop and the locomotive shed line. Trains could be signalled into either of the platform faces. The Exmouth distant signal was placed 960 yards from the box. Track circuiting appeared at the station in 1910 when it was installed between the down starting and down advanced starting signals. The opening of the extension to Budleigh Salterton on Whit Monday, 1 June 1903 brought a greater volume of traffic to the cramped layout and alterations in connection with the new work consisted of the island bay platform being extended, plus an additional set of facing points and new signalling and trackwork for the junction. The signal cabin was extended and the lever frame relocked, the frame then containing 30 levers, two being spare. Other track alterations took place in 1910 to ease the traffic working. Through coaches to and from Waterloo began in 1914, travelling via Tipton St John's and adding considerably to the problems of the congested station area.

The LSWR had a scheme for modernisation of the layout, and a plan was drafted in October 1916 that involved having a new station in the goods yard with four platforms of 600ft, and with the dock line that previously skirted around the goods shed now planned to run through the proposed station.

However, this was not to happen until the Grouping, when the Southern Railway completely rebuilt the station on an epic scale during 1924 at a total cost of £70,000 using the basis of the earlier LSWR plan. The original signalbox was removed and replaced by a 70-lever box standing between the running lines in the vee created by the convergence of the lines to Budleigh and Exeter. A new two-storey station building, constructed in red brick, was built behind the original building which was then demolished, this area being turned into a forecourt. With its glazed concourse, four platforms and a W. H. Smith bookstall, the whole layout had a commuter feel to it, even to the lattice gates to the platforms which were kept closed, being opened only 10 minutes before departures. In the 'mini Waterloo'-like atmosphere, the two island platforms, equipped with canopies and now numbered 1 to 4, could accommodate trains on either face. Two large four-armed gantries with LSWR pattern lower quadrant signals were positioned off the ends of both platforms, and trains could be signalled to either the Budleigh or Exeter lines from any of the platform faces. The new signalbox was equipped with two balconies from which the signalman could receive or hand over the single-line tablets. The line to Budleigh was known as the 'branch', and to Exeter the 'main line'. The branch and main inner home

Above: Adams 'O2' 0-4-4T No E187 awaits its turn at Exmouth on 18 June 1926 to leave with the 5.15pm to Tipton St John's. This locomotive was allocated to Yeovil Town in 1939 to replace 'D1' class tanks Nos 2299 and 2616 on the Town-Junction shuttles. *H. C. Casserley*

signals were fitted with electric route indicators, showing the platform arrival number to an incoming train, thus enabling the fireman to position himself in order to surrender the single-line token. During busy moments, of which there were plenty, when two trains were arriving or departing at the same time, the signalman would have his work cut out in handling the tokens for either line. Carriage sidings were also installed, the whole of the new layout coming into use on 20 July 1924. The goods sidings were remodelled and increased to seven, and a large brick-built goods shed was opened one day later on 21 July. The platform numbering was reversed from 2 June 1927 when No 1 became No 4 etc. The outside faces of the island platforms (1 and 4) had loops and were used for arriving trains, enabling locomotives to run round and shunt carriages into either Nos 3 or 2 thereby clearing the arrival platforms for the next inbound service. Staff consisted of a stationmaster, two booking clerks, two ticket collectors, two platform foremen, three porters, two signalmen and a porter-signalman, while, on the freight side, staff comprised a chief goods clerk, two goods clerks, two checkers, two goods porters, two shunters and three porter-guards. Trains to the docks, accompanied by a flagman, were propelled, running as required. When things were busy, two trains per day would run, coal being a

major import, and at one time all the town's coal was delivered by sea, with wagons also being despatched to Exeter Alphington Road, Exeter Central, Lympstone, East Budleigh and Budleigh Salterton. Other traffic imported included timber and wood pulp, and at one time there was a thriving pilchard industry. The heaviest locomotives permitted on the line were the Adams 'O2' locomotives.

'West Country' class No 21C115 *Exmouth* received her nameplates in a ceremony at the station on 26 June 1946, the locomotive having to travel via Sidmouth Junction and Tipton St John's to avoid the Clyst Viaduct weight restriction. The station was situated close to the town centre, but was a good walk from the better part of the sea front, many holidaymakers taking taxis rather than walking with their suitcases. As well as the summer Saturday through trains to and from Waterloo which ran via Tipton St John's, there was an unadvertised through train to Manchester departing from Exmouth at 9.18am. Upon arrival at Exeter Central, additional coaches are attached and a WR locomotive that had come up from Exeter St Davids, usually a 'Hall' 4-6-0, would take the train down the bank to St Davids and onwards over the ex-GWR route. Sunday excursions from Waterloo via Exeter, and holiday expresses from Bristol, Bridgwater and Plymouth also ran, with coaches filled to capacity.

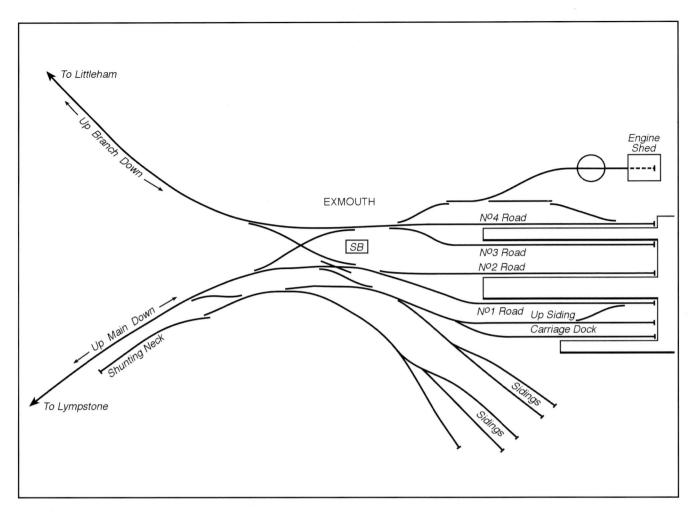

Below: Adams 'O2' No E182 runs round at Exmouth after arriving with the 1.30pm from Exeter on 18 June 1926. H. C. Casserley

Above: Drummond 'M7' 0-4-4T No 253 approaches Exmouth with a loose-coupled freight from Sidmouth Junction via Tipton St John's on 17 June 1949. *S. C. Nash*

The 1960s were not happy years for Exmouth; the locomotive shed closed on 8 November 1963 after the introduction of DMUs on most services, with closure of the line to Budleigh Salterton and Tipton St John's following on 6 March 1967. Freight facilities were withdrawn, including closure of the docks line, on 2 December, followed by the goods shed two days later. The signalbox closed on 10 March 1968 and all points were secured out of use, the line from Topsham being worked under the 'one engine in steam' regulations. The station was partially unstaffed from 6 May the same year, and all the redundant track was removed in 1969. The station was now a forlorn sight; all the track had been removed, except for the single line to Platform 4, and passengers were faced with the unwelcome sight of an uncared for and run-down station, with weeds and undergrowth sprouting on the unused platforms and along the deserted trackbed of the lifted lines. The viaduct that once carried the track from Budleigh

Salterton was demolished, as was the station building. Built by a railway company who cared for their passengers' requirements, it was destroyed by members of that unfortunate brand of managerial people who like to do away with buildings and services which are of use to the general community. A new transport interchange 'factory shed' type of building with a brick-built office and booking hall was erected in 1976, coming into public use on 2 May. The new construction now abuts the former Platform 2 which was reinstated to a shortened length of 481ft, with the single line slewed over from the former Platform 4. A relief road now occupies part of the site of the area of the former locomotive shed. The line is still buoyant with passenger traffic, although the short-sighted policy of reducing rail facilities at places such as Exmouth to a bus-stop-type terminus, with no hope of enlarging the trackwork for use by extra trains in future, thus taking cars off the road, must be a cause for concern.

'O2' class No S224 rumbles across Exmouth Viaduct with the 11.50am to Sidmouth, including a through coach for Waterloo, on 17 June 1949. Note how the viaduct was originally constructed to accommodate double track but never utilised as such. *S. C. Nash*

Above: An LSWR 13-ton open goods wagon, No S8170 in black livery, at Exmouth on 10 June 1963. The wagon is branded to work between Exmouth dock and Exmouth station only. *A. E. West*

Below: Adams Radial No 30582 stands at Exmouth with the Westward Television exhibition train on 11 March 1961. Westward Television was the then new commercial station for Cornwall, Devon and South Somerset. The train, which visited many stations in the operating area, contained information and exhibits regarding the new programmes etc, plus a TV studio. The television station eventually became Westcountry Television and now operates as part of Carlton TV. *S. C. Nash*

Above: BR Standard Class 3MT 2-6-2T No 82025 brings the 3.20pm from Budleigh Salterton across the 352yd Exmouth Viaduct on 12 October 1959. *R. C. Riley*

Below: Exmouth station in the late 1960s, showing the road vehicles of the era. The building still reflects its classic proportions and is bedecked with advertisements for Baker Engraving Company and the Exmouth Hire Centre. The iron railings have been removed and two of the brick pillars have also gone. A party of holidaymakers complete with suitcases await transport to their various hotels etc. An ugly municipal concrete lamp standard spoils the photographer's view of this grand building which, unfortunately, was demolished and replaced by a modest building of modern design which came into use on 2 May 1976. *Lens of Sutton*

Above: The 'mini Waterloo'-like interior of Exmouth station as returning holidaymakers proceed through the iron gates on to Platform 3 to board the train to Waterloo via Tipton St John's. *Lens of Sutton*

Below: The gates are shut to Platforms 1 and 2 at Exmouth station. The notice between the gates alludes to Exeter Central trains and reads, 'Front two coaches for St James' Park and Lympstone stations'. *Lens of Sutton*

Above: 'M7' class No 30323 with set No 154 at Exmouth awaits departure to Exeter Central on 12 October 1959.
R. C. Riley

Below: The generous proportions of the layout at Exmouth are seen to good effect as 'M7' No 30676 shunts stock on 13 October 1959. The locomotive shed can be seen to the left of the platform canopy. *R. C. Riley*

Above: Class 3MT 2-6-2T No 82011 approaches Exmouth with the 11.20am from Exeter Central on 15 June 1958. *S. C. Nash*

Below: BR Standard 2-6-2T No 82024 has just run past the down branch outer home signal on the approaches to Exmouth with the 10.50am from Budleigh Salterton on 12 August 1960. *Terry Gough*

Class 3MT 2-6-2T No 82017 pulls away from Exmouth with the
1.45pm to Exeter Central on 13 October 1959.
R. C. Riley

The signalman can be seen on the box veranda waiting to pass the single-line token to the fireman of No 82017 leaving Exmouth with the 1.45pm to Exeter Central on 13 October 1959. The line to Tipton St John's is visible curving away to the right.
R. C. Riley

Left: The Exmouth home signals had indicators fitted showing the platform arrival number to footplate crews. This is the inner home signal (from Topsham) guiding an arriving train into Platform 1. *A. E. West*

Below: The waters of the Exe Estuary lap gently against the railway embankment at Exmouth as 2-6-2T No 82024 arrives with the 12.15pm from Exeter Central on 12 August 1960. *Terry Gough*

Right: A driver's-eye view of the buffer stops and water columns to Platforms 2 and 3 at Exmouth on 10 June 1963. Note how the gates are closed to the platforms in true Waterloo style. *A. E. West*

Exmouth Locomotive Shed

The first locomotive shed, dating from the opening of the line, was constructed of wood and comprised a single-ended building complete with an inspection pit, situated to the east of the station, close to New Street. A turntable was not provided at first, but a 42ft turntable was ultimately installed and positioned immediately outside the building. A water tank and an extra inspection pit were provided in the yard. The original water tank was found to be lacking in capacity, and a new tank was supplied in 1871 at a cost of £40. The two tanks then stood side by side until the opening of the new shed in later years. A mess hut was also positioned in the shed yard. Locomotives received heavy repairs and boiler washouts etc at the main locomotive shed at Exeter Queen Street, and ultimately, Exmouth Junction when opened in 1887. The old locomotive shed was replaced by a 70ft x 20ft concrete version supplied from the works at Exmouth Junction in 1927 at a cost of £2,200. The turntable, which was still extant but had been out of use for some years, was removed. A large water tank was erected at the rear of the new shed, which was also equipped with a small office, while a messroom, store and a WC, an

inspection pit and water hydrant were provided inside the building. Two further inspection pits were placed in the shed yard. Coaling for topping up the bunkers was accomplished from wagons stabled on an adjacent siding but coaling was mostly dealt with at Exmouth Junction, with time allotted in various diagrams etc. By the mid-1920s, staff included eight pairs of enginemen, supported by approximately three engine cleaners who attended to the stabled engines' needs. A spare gang from Exmouth Junction assisted with the heavy summer traffic. The 'O2s' and 'M7s' were normally stabled here, interspersing with the 2-6-2 tanks when they arrived in the 1950s.

The Beattie 2-2-2 well tanks were at work on the line from the start, being superseded by standard 2-4-0 well tanks from 1878. An 'Ilfracombe Goods' 0-6-0, No 301, was allocated to Exmouth in June 1881. One of Joseph Beattie's 'Volcano' class 2-4-0 tender engines, No 11 (named *Minerva* in September 1883), was allocated to Exmouth in March 1878, this locomotive being the only member of her class stationed in the West Country. These locomotives worked all the Exmouth passenger services, starting with the 7.15am Exmouth-Exeter and finishing with the 10.25pm from Exeter. 'Saxon' class 2-4-0 No 137 *Hun*, and 'Gem' class 2-4-0 *Meteor* shared the daily 1.10pm down and 2.40pm return Exmouth goods in 1878; later this duty was worked by a

Above: No 30667 stands next to the ash truck at Exmouth shed on 29 August 1957. As No 667, this was one of the original batch of 25 short-frame 'M7s' built at Nine Elms in August 1897. It was converted to air-operated push-and-pull in March 1961 using the long frames and motor train gear from No 30128, and was withdrawn in May 1964. *A. E. West*

Beyer Peacock double-framed 0-6-0 goods No 0288, which in turn was displaced by 'Ilfracombe Goods' Nos 0283, 0284, and 0393 when allocated to Exmouth Junction in the early 1900s. Adams '0380' class 4-4-0s Nos 0381, 0383 and 0384 were based at Exmouth in 1905 for the goods services. 'O2' tanks Nos 179 and 187 appeared at Exmouth Junction in March 1900 for the Exmouth services, followed by Nos 209 and 231 which worked the services in company with Adams '0415' class 4-4-2 tanks Nos 0520, 0523 and 0107. Up to and including 1932, most of the branch operation on the Exeter-Exmouth-Sidmouth services was in the hands of the 'O2' tanks, of which Exmouth Junction had a plentiful supply, comprising Nos 178, 181, 191, 198, 203, 207, 208, 214, 228, 231, 232 and 235. However, the restriction on 'M7s' traversing the Topsham-Exmouth section was lifted in July 1932, thus allowing the Drummond tanks to flood on to the line (although they had been able to travel to Exmouth via Sidmouth Junction), thereby releasing some of the 'O2s' to be withdrawn, although they were at work on the line almost to the end of steam. With 'M7s' being too heavy to be used on the Topsham Quay and Exmouth Docks lines, these remained the preserve of the 'O2s'. The line to Exmouth Docks was worked in daylight hours only, the trains being subject to a 4mph speed restriction and accompanied by a man with a red flag. A Manning Wardle 0-6-0T, No 458 *Jumbo,* was allocated to Exmouth Junction in 1895 for the Exmouth docks line previously the locomotive had been working on the Bodmin & Wadebridge Railway.

Ivatt Class 2MT 2-6-2 tanks Nos 41313-5 and BR Class 3MT 2-6-2 tanks Nos 82010-3 were delivered to the Southern Region and allocated to Exmouth Junction in 1952 for use on the Exmouth, Sidmouth Junction-Sidmouth, Honiton, Axminster and Crediton services. No 41314 performed a series of trials between Exeter and Exmouth on 22 July 1952, hauling the maximum 'M7' loading of seven coaches, and although timekeeping was exemplary, the locomotive did not prove better than the 'M7'. A superior effort was demonstrated by No 82011, and by October 1952 Nos 82010-3/7-9 were monopolising the services with occasional help from No 41314 and 'M7s' Nos 30023/5/45, 30667/9. The change from the old 'M7s' was a vast improvement for the locomotive crews; the Standard tanks had more roomy and comfortable cabs, free steaming

234

Above: 'M7' No 30025 stands in the shed yard at Exmouth in company with a 2-6-2T in 1958. *John Day*

boilers, and could handle the Exmouth suburban services with ease. Standard Class 2 2-6-2Ts Nos 84020-3 were allocated to Exmouth Junction for the branch services in May 1961, but not for long as they were sent off to the London Midland Region by the early autumn of the same year. The final appearance of an 'M7' on the line occurred with No 30025 in April 1963. The first of the powerful Class 4 2-6-4 Standard tanks, No 80035 was allocated to Exmouth Junction in June 1962; the whole contingent, when completed some months later, comprising Nos 80035-43/59/64/67, were utilised on the branch services.

Dieselisation occurred in 1963, the last steam passenger trains being operated on 4 January 1965 involving the 8.20am from Exmouth and the 5.45pm from Exeter, the final steam-hauled freight working being on 24 May the same year. The locomotive diagrams used on the Exeter-Exmouth-Sidmouth-Sidmouth Junction passenger and freight services in 1963 consisted of Exmouth Junction Duty Nos 608, 609, 610, 612, 613, 614, 615, 616, 617, 626, 627, 628 and 629. An interesting feature of Duty No 617 was that upon arrival at Exeter Central with the 1.45pm from Exmouth, the locomotive was diagrammed to bank a stone train from St Davids before resuming work on the Exmouth line with the 3.18pm from Exeter Central. Some fast work was required, as the stone train, if running to time, was not booked into Exeter Central until 3.13pm.

Above: Class 3MT 2-6-2T No 82025 arrives at Exmouth with the 12.45pm from Exeter Central on 12 August 1960. The superb LSWR lower quadrant signal arms were always a feature of this station. *Terry Gough*

Footplate Days at Yeovil Town

I always look back on my engine cleaning days at Yeovil Town with great fondness. Working in the shed amongst the noise and bustle of hammering and banging as the fitters strove away at their work, water cascading down into the pits as a boiler was being washed out, shouting and whistling — all of this was tempered with the noises outside of safety valves lifting and the steam crane hoisting tubs of coal on to almost empty tenders and locomotives moving with a roaring of steam from open cylinder cocks. But engine cleaning and the myriad duties that came with it, from delivering call outs to the off-duty enginemen, to looking after the sand furnace, ensuring that ample supplies of dry sand were available for locomotives, do not last for ever.

Starting as I did on the footplate grades as an engine cleaner, the next step on the ladder was 'passed engine cleaner', before rising to the rank of fireman. On the Western Section of the Southern Region, this meant firing a passenger train from Exeter Central to Yeovil Junction under the eagle eye of Sam Smith, the Locomotive Inspector at Exmouth Junction. Everybody on the line knew him; he was a formidable-looking figure and, up until then, I had not personally come into contact with him, although I had seen him at Yeovil Town shed on various occasions. Having previously been passed out at the Exeter firing school on Rules and Regulations etc, it was my turn to pass the firing test and turn theory into practice. The usual way of things, which had been done by scores of engine cleaners from Yeovil over the years, was to travel to Exeter on the footplate of a down train from Yeovil Junction with a Yeovil crew. This was done with the crew, whom I knew, allowing me to attend to the locomotive under the guidance of the fireman. All went well, with plenty of information given to me, including one of the most important: 'do not soak Sam Smith's boots when you wash down the footplate with the pep pipe.' All well and good. Most engine cleaners would be apprehensive at the very least, and I was no exception as we neared Exeter. To cut a long story short, we were eventually relieved and I now stood at the end of the up platform at Exeter Central. The station was very busy: main line trains arriving and departing with much splitting and rejoining, branch trains running in and out almost nonstop, Drummond

tanks shunting in the goods and carriage sidings and the carriage cleaning shed, light engines running to and fro, and the platforms very busy with passengers. My driver was chatting away to Sam Smith while the fireman and I had a cigarette by the water column. Our train was coming from Plymouth, with the locomotive working through to Yeovil Junction. The Inspector had already checked my name and number etc, and passengers were gathering on the platform for our train. The 'boards' were off, this meaning that the train was coming up the bank. No time for butterflies now; the station tannoy crackled into life announcing the arrival of the Plymouth-Salisbury stopper. By now I could hear that well-known, quiet chuffing of a Bulleid Pacific as she swung into the platform line and clanked along towards us, the sound of the vacuum exhaust and the roaring of steam growing louder by the second. The track trembled under the immense weight of the locomotive and carriages. As soon as the engine had stopped, I hopped up onto the footplate, put my bag away and, like a monkey, having opened the tender doors, scampered over the coal and opened the filler flap. The column had already been dragged around, the fireman turning the water on as soon as the bag went in. In no time at all we had filled the tender, and back I went to the footplate. The Plymouth crew were by this time talking to the Inspector on the platform; it was rumoured that Sam knew every driver and fireman between Salisbury and Wadebridge. This I believed.

The fireman whom I had come down with had entered the train, so I was now on my own. The footplate was so hot it took my breath away. The valves were starting to lift; there were three parts of a glass of water in the gauges, and the vacuum gauge was showing the correct 21 inches. The large ejector hissed and spat and the whole locomotive was alive and ready to go. A huge fire burnt away, scorching my overalls through the open firedoor. The Inspector climbed aboard and clipped the door shut. My driver glanced back at our train, and the station staff closed the last few doors. A long whistle blast from the guard, accompanied by a flurry from his green flag, and we were off. My mate opened the regulator, and with a quiet chuff from the exhaust we began to move forward, and in accordance with the Rules I looked to the rear of our train along the length of the platform just in case some fool tried to run and catch the train at the last minute. I have seen it happen, and more than once a late passenger has nearly stumbled down the gap 'twixt platform and moving train'. It was now time to close the firedoor and stand at my side of the footplate. The locomotive pulled away with more speed as we clattered over the points and crossings past the lines of dark green coaches away to our right outside the carriage cleaning shed. A passenger train inbound from Exmouth rumbled past; more regulator was applied and the reverser eased back. The locomotive was well into her stride, the Inspector watched me as I opened the firedoor and deflected the heat with my firing shovel to see where the fire needed attention. The maxim with the Bulleid Pacifics was, 'keep the back corners full'. The 'West Countrys' have a steam-operated firedoor operated by a foot pedal, the knack being to load your shovel from the tender, turn and at the same time press your foot on the pedal. The firedoors open, you throw in the coal, release the pedal and the doors shut — quite a good system. I sensed the Inspector watching me from the corner of his eye as I packed some rounds of coal around the back corners as I had been taught. The pressure gauge was holding as we sprinted up the incline towards Blackboy Tunnel. There was a loud ring from the AWS bell as we caught the Exmouth Junction colour-light distant signal showing a green aspect. There was no more need to fire at this moment; the water was dropping in the gauges, but this was not the time to put on the steam injector — that could wait until we were running past Exmouth Junction shed, to keep the engine quiet. The Stones generator had been humming away, with all electric lights blazing, since we had relieved the Plymouth crew. The gaping tunnel mouth lay dead ahead with wisps of steam drifting into the atmosphere as we ran past St James' Park Halt. With a blast on the whistle we ran into the dark interior. Our exhaust blast rebounded from the tunnel walls, the lights from our carriages reflected in the darkness, red hot clinker from our ashpan dropped on to the sleepers. Then we were out into the open and bounding up the last few yards of the incline. Scores of wagons stood in Exmouth Junction yard to our left; all the 'boards' were off as we hurtled over the junction with the Exmouth branch streaking away to our right. The sound of lifting safety valves resounded from Exmouth Junction depot as we ran past.

The Pinhoe distant was off, and the AWS bell rang again in the cab as my mate shut off and we coasted into the station. Our train was booked to stop at every station except Sutton Bingham, so I just got on with the job, firing to the locomotive when needed, keeping the boiler topped up and adhering to the Rule Book when not tending to the fire, keeping the footplate clean and tidy. And I did not splash the Inspector's boots with the pep pipe! The Inspector now stood beside my driver. They were chatting away merrily, so he must have been satisfied with my efforts so far. By the time we had reached Sidmouth Junction my anxieties as to the firing test had vanished. Sam Smith seemed unconcerned and left me to get on with it.

The Yeovil Junction distant was showing caution as we approached and I had been running the fire down slowly as we were due to come off there, with a Yeovil 'U' class 2-6-0 taking the train on to Salisbury. The regulator was shut with just a turn on the steam blower to stop a blow-back from the firebox. The outer and inner home signals were off as we approached and swung into the up platform line. My mate brought the large ejector handle down with an almighty hiss as we clanked along the platform, bringing the train to a halt at the end of the platform just in front of the six-arm gantry. Armed with a headlamp to place on our tender as a tail lamp, I climbed down from the cab and swiftly uncoupled from our train. The up local to down branch signal was off, and we quickly headed down the branch to Yeovil Town, leaving our mates on the 'U' class to reverse down from the signalbox and take the train onwards. Upon arrival at Yeovil Town Loco, our locomotive was screwed down on the pit, and while my mate walked to the drivers' cabin, the Inspector and I wandered over to a locomotive standing on the shed sidings. Here, I had to name every piece of the locomotive valve gear and what it does in relation to the locomotive itself. After that it was question and answer time on Rules and Regulations. When this was completed, Sam Smith told me that I had passed — and that was it.

As the Inspector wandered towards the shed office for a cup of tea, I joined my mates in the cabin for a well-earned Woodbine! Promotion to fireman would come later but, for now, I was looking forward to being booked out for firing duties on the main line. Ahead of me lay firing days with 'T9s', 'Black Motors', 'M7' and 'O2' tanks, 'Rivers' and 'Woolworths', 'Blackuns' and 'Arthurs', 'West Countrys' and 'Packets'. Rough trips intermingled with the good, and the almost Masonic lifestyle of the steam footplatemen of which I was proud to be a member. For we shall never see the like again.

Index

'M7' 0-4-4T No 30055 in early British Railways livery working an up train near Colyford on 17 June 1949. *S. C. Nash*